General Physics

PEARLS OF WISDOM

DAVID AMSTUTZ, Ph.D.
University of Nebraska at Kearney

> **NOTE**
>
> The intent of General Physics Pearls of Wisdom is to serve as a study aid to improve performance on a standardized examination. Neither Boston Medical Publishing Corporation nor the editors warrant that the information in this text is complete or accurate. The reader is encouraged to verify each answer in several references.

Copyright © 2001 by Boston Medical Publishing Corporation, Lincoln, NE.

Printed in U.S.A.

All rights reserved, including the right of reproduction, in whole or in part, in any form.

The editors would like to extend thanks to Terri Lair for her excellent managing and editorial support.

Special thanks to Donald Schlegel, Professor of Sciences at College View Academy in Lincoln, Nebraska, for his review of this book.

Art Director Maryse Charette

This book was produced using Times, GillSans and Symbols fonts and computer based graphics with Macintosh® computers

ISBN: 1-890369-23-3

DEDICATION

*This volume is dedicated to Polly and David Lee,
who make life worthwhile.*

David Amstutz

WE APPRECIATE YOUR COMMENTS!

We appreciate your opinion and encourage you to send us any suggestions or recommendations. Please let us know if you discover any errors, or if there is any way we can make Pearls of Wisdom more helpful to you. We are also interested in recruiting new authors and editors. Please call, write, fax or e-mail. We look forward to hearing from you.

Return to:

Boston Medical Publishing Corporation
4780 Linden Street, Lincoln, NE, 68516

888-MBOARDS (626-2737)
402-484-6118
Fax: 402-484-6552
E-mail: bmp@emedicine.com
www.bmppearls.com

INTRODUCTION

Congratulations! General Physics Pearls of Wisdom will help you improve your knowledge base in physics. A few words are appropriate in discussing intent, format, limitations and use.

Since Pearls is primarily intended as a study aid, the text is written in a rapid-fire question/answer format. This way, readers receive immediate gratification. Moreover, misleading or confusing "foils" are not provided. This eliminates the risk of erroneously assimilating an incorrect piece of information that makes a big impression. Questions themselves often contain a "pearl" intended to reinforce the answer. Additional "hooks" may be attached to the answer in various forms, including mnemonics, visual imagery, repetition, and humor. Additional information, not requested in the question, may be included in the answer. Emphasis has been placed on distilling trivia and key facts that are easily overlooked, quickly forgotten and that somehow seem to be needed on MCAT, DAT, OAT and similar examinations.

Many questions have answers without explanations. This enhances ease of reading and rate of learning. Explanations often occur in a later question/answer. Upon reading an answer, the reader may think, "Hm, why is that?" or "Are you sure?" If this happens to you, go check! Truly assimilating these disparate facts into a framework of knowledge absolutely requires further reading of the surrounding concepts. Information learned in response to seeking an answer to a particular question is retained much better than information that is passively observed. Take advantage of this! Use Pearls with your preferred texts handy and open.

Pearls has limitations. We have found many conflicts between sources of information. We have tried to verify in several references the most accurate information. Some texts have internal discrepancies further confounding clarification.

Pearls risks accuracy by aggressively pruning complex concepts down to the simplest kernel—the dynamic knowledge base and practice of Physics is not like that! Furthermore, new research and practice occasionally deviates from that which likely represents the right answer for test purposes. This text is designed to maximize your score on a test. Refer to your most current sources of information and mentors for further direction.

Pearls is designed to be used, not just read. It is an interactive text. Use a 3 x 5 card and cover the answers; attempt all questions. A study method we recommend is oral, group study, preferably over an extended meal or pitchers. The mechanics of this method are simple and no one ever appears stupid. One person holds Pearls, with answers covered and reads the question. Each person, including the reader, says "Check!" when he or she has an answer in mind. After everyone has "checked" in, someone states his/her answer. If this answer is correct, on to the next one; if not, another person says their answer or the answer can be read. Usually the person who "checks" in first receives the first shot at stating the answer. If this person is being a smarty-pants answer-hog, then others can take turns. Try it, it's almost fun!

Pearls is also designed to be re-used several times to allow, dare we use the word, memorization. Two check boxes are provided for any scheme of keeping track of questions answered correctly or incorrectly.

We welcome your comments, suggestions and criticism. Great effort has been made to verify these questions and answers. Some answers may not be the answer you would prefer. Most often this is attributable to variance between original sources. Please make us aware of any errors you find. We hope to make continuous improvements and would greatly appreciate any input with regard to format, organization, content, and presentation or about specific questions. We also are interested in recruiting new contributing authors and publishing new textbooks. Contact our manager at Boston Medical Publishing, Terri Lair, at (toll free) 1-888-MBOARDS. We look forward to hearing from you!

Study hard and good luck!

D.C.

TABLE OF CONTENTS

MECHANICS
- UNITS CONVERSIONS AND VECTORS ... 9
- MOTION IN ONE DIMENSION ... 17
- TWO-DIMENSIONAL MOTION AND RELATIVE VELOCITY ... 27
- NEWTON'S LAW AND FRICTION ... 41
- NEWTON'S LAW II ... 53
- WORK, POWER AND ENERGY ... 63
- IMPULSE, MOMENTUM AND COLLISIONS ... 73
- RATIONAL MOTION I ... 83
- RATIONAL MOTION II ... 89
- ELASTICITY AND SIMPLE HARMONIC MOTION ... 101
- FLUIDS ... 111

HEAT AND THERMODYNAMICS
- TEMPERATURE SCALES AND EXPANSION ... 123
- HEAT ... 131
- HEAT TRANSFER ... 137
- THERMODYNAMICS ... 145

WAVES AND SOUND
- THE MOTION OF WAVES ... 155
- SOUND ... 163

ELECTRICITY AND MAGNETISM
- CHARGES AT REST ... 171
- POTENTIAL ... 183
- CAPACITORS ... 191
- CURRENT, RESISTANCE AND SOURCES OF EMF ... 201
- DIRECT CURRENT CIRCUITS ... 211
- MAGNETIC FIELDS AND FORCES ... 223
- FARADAW'S LAW AND ELECTRO-MAGNETIC INDUCTION ... 235
- ALERNATING CURRENT AND TRANSFORMERS ... 249

OPTICS
- WAVES AND LIGHT ... 255
- GEOMETRIC OPTICS I ... 261
- GEOMETRIC OPTICS II ... 275
- WAVE OPTICS I ... 291
- WAVE OPTICS II ... 303

GENERAL PHYSICS PEARLS OF WISDOM

MECHANICS
UNITS, UNITS CONVERSIONS, AND VECTORS

THE FUNDAMENTAL IDEAS ARE:

- To convert units: just multiply by the well-chosen one. To do this, you will need conversion facts, such as 1 mile = 1609 m, 1 min = 60 s, 1 km/hr = .2778 m/s.

- A vector is any quantity that has magnitude and direction. this means that it takes magnitude and direction to specify a vector.

- A scalar is any quantity that has magnitude only.

- To find the components of a vector, one uses one of the trigonometric functions: sine, cosine, or tangent.

- To find a vector given its components, use the Pythagorean Theorem to find its magnitude and a trig function to find its direction. Note that vectors are usually specified using the geographic (north, east, south, west) or mathematical (x,y) frame of reference.

- To add vectors together, just add corresponding components together. A vector addition table is helpful here.

- Note that vectors are denoted by **bold-faced** type and that scalars and magnitudes of vectors are denoted by regular type.

SOLVED PROBLEMS

▫▫ The distance from A to B is 96 miles. How far is that in meters? [1 mile = 1609 m]

$$96 \text{ miles} \left(\frac{1609 m}{1 mile}\right) \cong 154{,}000 \text{ m}$$

Note that all units cancel except m.

▫▫ A large diamond has a size of 2500 carats. What does the diamond weigh in pounds?
[1 carat = .2g; 1 kg = 2.2 pounds for the purposes of this problem]

$$2500 \text{ carat} \left(\frac{.2g}{1 carat}\right)\left(\frac{1 kg}{1000 g}\right)\left(\frac{2.2 lbs}{1 kg}\right) \cong 1.1 \text{ lbs}$$

All units cancel except pounds.

□□ **How many seconds are there in a day?**

$$1 \text{ day} \left(\frac{24 hr}{1 day}\right)\left(\frac{60 \min}{1 hr}\right)\left(\frac{60 s}{1 \min}\right) \cong 86,400 \text{ s}$$

All units cancel except seconds.

□□ **An aircraft has a speed of 600 knots. What is its speed in miles per hour? [1 knot = 1 nautical mile/hour; 1 nautical mile = 6076 feet; 1 mile = 5280 ft]**

$$600 \text{ knots} = \left(\frac{600 \text{ nautical miles}}{hour}\right)\left(\frac{6076 \text{ feet}}{1 \text{ nautical mile}}\right)\left(\frac{1 \text{ mile}}{5280 \text{ feet}}\right) \cong 690 \text{ miles/hour}$$

□□ **The 484 pages of a book are 3.9 cm thick. find the average thickness of a single sheet of paper used in the book. Express your answer in microns, μ. [1 cm = 1 x 10^{-2}m; 1 μ = 1 x 10^{-6} m]**

Since 484 pages = 242 sheets, we have

242 sheets = 3.9 cm and thus 1 sheet = $\left(\frac{3.9 cm}{242}\right)\left(\frac{1 \times 10^{-2} m}{1 cm}\right)\left(\frac{1\mu}{1 \times 10^{-6} m}\right) \cong 161\mu$

□□ **Find the components of vector A in the adjacent figure:**

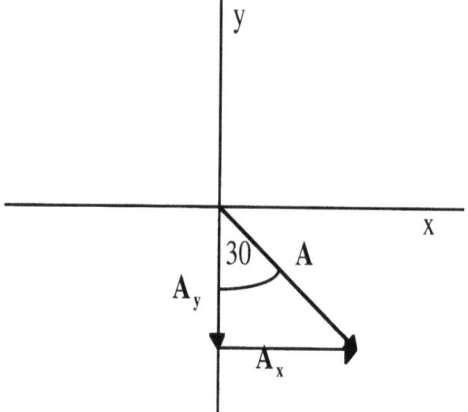

The magnitude of **A** is 10 km.

$A_x = A \sin 30°$
    $= 5$ km
$A_y = A \cos 30°$
    $= -8.66$ km

☐☐ **Find the components of vector B in the adjacent figure:**

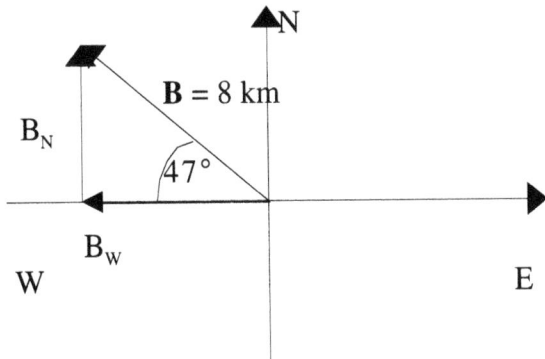

B_N = B sin 47°
= 5.85 km

B_W = B cos 47°
= 5.46 km

☐☐ **Find vector R in the adjacent figure:**

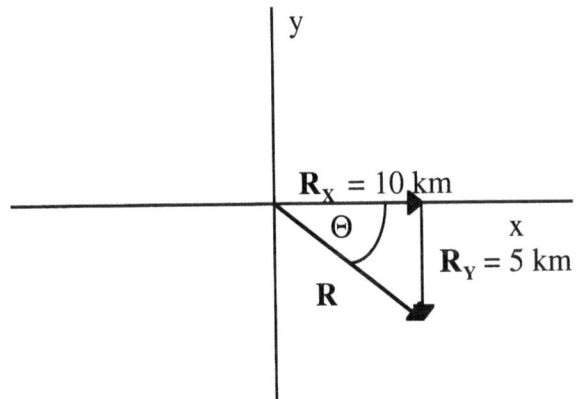

$R = \sqrt{R_X^2 + R_Y^2} \cong 11.2 \; km$

$\tan \Theta = \dfrac{R_Y}{R_X} = \dfrac{5\,km}{10\,km} = .5$

$\Theta = 26.6°$
Vector **R** = 11.2 km @ 26.6° below + x axis or @ 333°.

☐☐ Find A in the given figure:

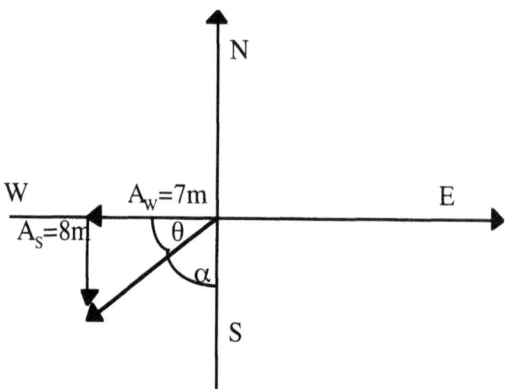

$A = \sqrt{A_S^2 + A_W^2} = \sqrt{(7m)^2 + (8m)^2}$

$A = 10.6$ m

$\tan \theta = \dfrac{A_S}{A_W} = \dfrac{8m}{7m} = 1.143$

$\theta = 48.8°$
⇒ **A** ≅ 10.6 m @ 48.8° S of W
or **A** ≅ 10.6 m @ 41.2° W of S (angle α in the figure)

☐☐ An ant crawls 5 m west, 8 m south and 6 m 30° N of W. What is the ant's displacement from its starting position?

vector	E	N
a	-5m	0
b	0	-8m
c	-5.2m	3m
R	-10.2m	5m

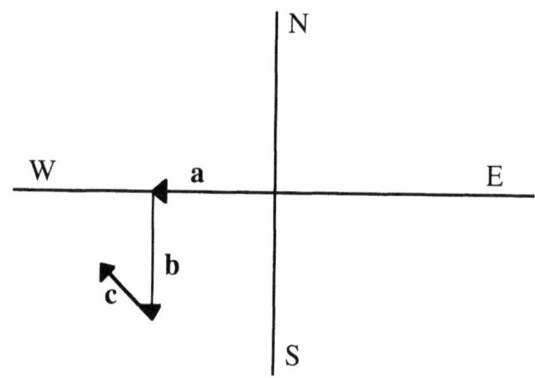

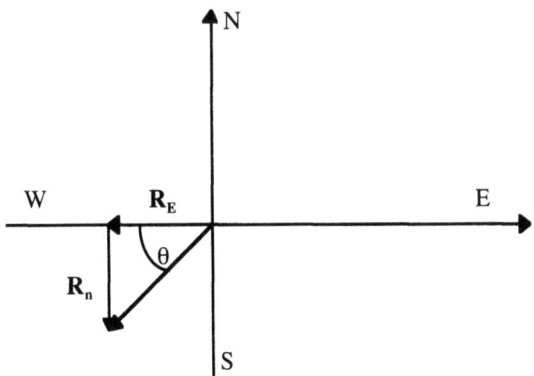

$R = \sqrt{E^2 + N^2}$

$\Rightarrow R = \sqrt{(-10.2m)^2 + (5m)^2} \approx 11.4m$

$\tan \theta = \dfrac{R_N}{R_E} = \dfrac{5m}{10.2m} = 0.49$

$\Rightarrow \theta = 26.2°$

R ≅ 11.4 m @ 26.1° S of W   or

R ≅ 11.4 m @ 63.9° W of S

☐☐ **What is the resultant of the three vectors in the given figure?**

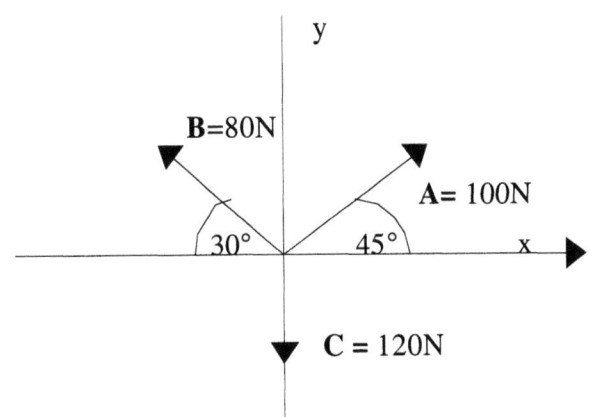

vector	x	y
A	70.7 N	70.7 N
B	-69.3 N	40 N
C	0	-120 N
R	1.4 N	-9.3 N

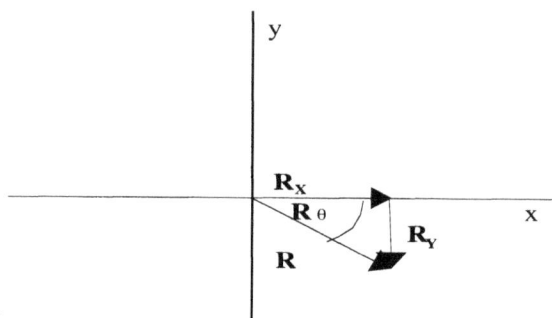

$A_x = A \cos 45° = 70.7$ N
$A_y = A \sin 45° = 70.7$ N

$B_x = B \cos 30° = 69.3$ N
$B_y = B \sin 30° = 40$ N

$R = \sqrt{x^2 + y^2} = \sqrt{(1.4N)^2 + (-9.3N)^2}$

$\cong 9.4$ N

$\tan \theta = \dfrac{Ry}{Rx} = \dfrac{9.3N}{1.4N} = 6.64$

$\theta = 81.4°$ so

R ≈ 9.4 N @ $81°$ below + x axis.

In astronomy distances are so large that several different units are used. one is the astronomical unit (AU) which is 1.49×10^{11} m. Another unit is the light year, which is the distance light travels in one year, and a third is the parsec, the distance at which one AU subtends an angle of one second of arc.

❏❏ **If light travels at 3×10^8 meters per second, how many kilometers are in a light year?**

$3 \times 10^8 \dfrac{m}{s} \left(\dfrac{86,400s}{day} \right) \left(\dfrac{365 days}{year} \right) \approx 9.46 \times 10^{15} \dfrac{m}{year}$ (note that 1 km = 1000 m)

Thus, light travels 9.46×10^{12} km/year

❏❏ **How many AU are in a light year?**

1 light year = 9.46×10^{12} km $\dfrac{1AU}{1.49 \times 10^8 km} \cong 6.35 \times 10^4$ AU

❏❏ **How many AU are in a parsec?**

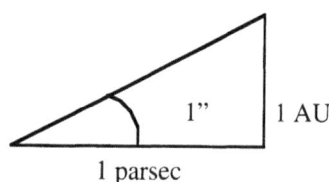

$$\theta(rad) = \frac{S}{R} \Rightarrow R = \frac{S}{\theta(rad)} = \frac{1 AU}{(1'')\left(\frac{1'}{60''}\right)\left(\frac{1°}{60'}\right)\left(\frac{1 rad}{57.3°}\right)}$$

$R \approx 206{,}000$ AU   so that,

1 parsec $\cong 206{,}000$ AU

☐☐ **An ultra-light flies 18 km E, 14 km 20° S of E, and 9 km W. What displacement (magnitude and direction) is needed for the ultra-light to return to its starting point?**

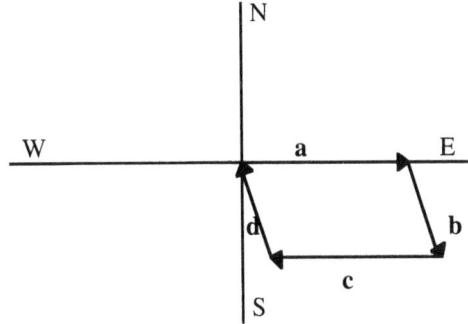

vector	E	N
a	18km	0
b	13.2km	-4.8km
c	-9km	0
d	d_E	d_N
R	0	0

$b_E = b \cos 20° \cong 13.2$ km

$b_N = b \sin 20° \cong -4.8$ km

One adds corresponding components to add vectors together, and when the ultra-light is back where it started from its displacement is zero.

East component: $0 = 18km + 13.2km - 9km + d_E \Rightarrow d_E = -22.2$ km

North component: $0 = 0 - 4.8km + 0 + d_N \Rightarrow d_N = 4.8$ km

$d = \sqrt{N^2 + E^2} = \sqrt{(4.8km)^2 + (-22.2km)^2}$
$= 22.7$ km

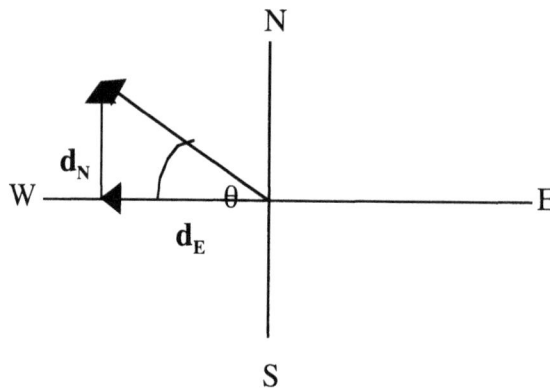

$$\tan\theta = \frac{d_N}{d_E} = \frac{4.8km}{22.2km} = .216$$

$\Rightarrow \theta = 12.2°$

so d = 22.7km @ 12.2° N of W

MOTION IN ONE DIMENSION

THE FUNDAMENTAL IDEAS ARE:

- The change in any quantity is the difference between the initial and final value, i.e. $\Delta Q = Q - Q_o$ where Δ means change and the subscript $_o$ means original or initial value and lack of a subscript denotes a final value.

- Displacement is a vector whose magnitude is distance. Units of displacement are m, km, cm, etc.

- $\Delta x = x - x_o$

- Average velocity, $v_{av} = \dfrac{\Delta x}{\Delta t}$. Note that the units of velocity are units of displacement divided by time, e.g., m/s, miles/hour, cm/year.

- Average acceleration, $a_{av} = \dfrac{\Delta v}{\Delta t}$. The units of acceleration are units of velocity divided by time, e.g., m/s/s or m/s^2 , miles/hr/hr or mi/hr^2, ft/s/s or ft/s^2.

- Distance is magnitude of displacement, average speed is distance divided by time, instantaneous speed is the magnitude of the instantaneous velocity (e.g., the reading of a speedometer).

- The five tools of motion are:
  (1) $x = v_{av} t$
  (2) $v_{av} = \dfrac{v + v_0}{2}$
  (3) $v = v_o + at$
  (4) $v^2 = v_o^2 + 2ax$
  (5) $x = \dfrac{1}{2} at^2 + v_0 t$

- The tools of motion fail if the acceleration is not constant.

- If the acceleration is not constant, the tools of motion can still be used if the motion is broken up into segments where the acceleration is constant.

- Any object that is in free-fall near the surface of the earth experiences an acceleration directed in the down sense that has magnitude of 9.8 m/s^2 or 32 ft/s^2 and is called the acceleration due to gravity.

- Displacement, velocity and acceleration are vectors. The usual convention regarding sign is as indicated in the figure.

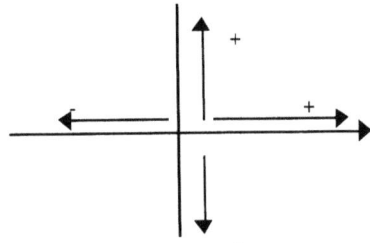

SOLVED PROBLEMS

❑❑ **A car travels west from Kearney, Ne (mile marker 273 on I-80) to North Platte, Ne (mile marker 177 on I-80). What are the displacement and the distance that the car travels?**

(a) The displacement is $\Delta x = x - x_o$
$= 177 \text{ miles} - 273 \text{ miles}$
$= -96 \text{ miles}$
thus, $x = 96$ miles west

(b) The distance (magnitude of displacement) is 96 miles

❑❑ **If the car in the previous problem takes 1.5 hours to travel to North Platte, what was (a) the average velocity and (b) the average speed of the car?**

(a) $v_{av} = \dfrac{\Delta x}{\Delta t} = \dfrac{96 \text{miles}}{1.5 \text{hours}} \text{ west} = 64 \dfrac{\text{miles}}{\text{hour}} \text{ west}$

(b) $(\text{speed})_{av} = \dfrac{96 \text{miles}}{1.5 \text{hours}} = 64 \dfrac{\text{miles}}{\text{hour}}$

❑❑ **An object is thrown up with an initial velocity of 7 m/s. Find (a) how high the object rises, (b) how long the object is in the air, and (c) its velocity when it returns to the same level from which it was thrown.**

(a)

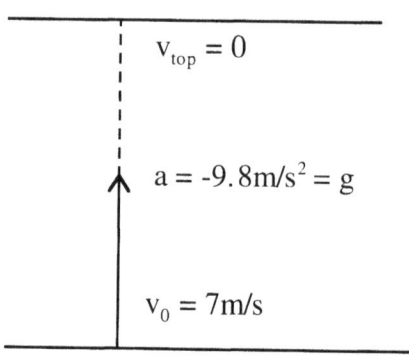

from tool (3),
$v = v_0 + at$

$0 = 7 \text{ m/s} + (-9.8 \text{ m/s}^2)t$
$\Rightarrow t = .714$ s to reach maximum height
$\Rightarrow$ time in air $= 2(.714) \approx 1.43$ s, since $t_{up} = t_{down}$

(b) Now $y_{max} = v_{av} t$ (using tool 1), so that

$y_{max} = \left(\dfrac{7 m/s + 0}{2} \right)(.714s) = 2.5 m$

Second Method (using tool 5)

$$y_{max} = \frac{1}{2}at^2 + v_0 t$$

$$= \frac{1}{2}\left(-9.8\frac{m}{s^2}\right)(.714s)^2 + 7\frac{m}{s}(.714s)$$

$$= 2.5 \text{ m}$$

(c) (using tool 4) $v^2 = v_0^2 + 2ax$
$\Rightarrow v^2 = 49 \text{ m}^2/\text{s}^2 + (-9.8 \text{m/s}^2)(0)$
$v = \pm 7 \text{ m/s}$
$v = -7 \text{ m/s}$, because the object is going down

(using tool 3) $v = v_0 + at$
$= 7 \text{ m/s} - 9.8 \text{ m/s}^2(1.43)$
$= -7 \text{ m/s}$

☐☐ **An object is moving with constant acceleration. At t = 0, its velocity is 9 m/s south. 5 seconds later its velocity is 7 m/s north. Find its acceleration.**

(using tool 3) $v = v_0 + at \Rightarrow a = \dfrac{v - v_0}{a}$

$$= \dfrac{7\frac{m}{s} - \left(-9\frac{m}{s}\right)}{5s}$$

$a = 3.2 \text{ m/s}^2$ north.

☐☐ **A train starts from rest and travels 245 m in 34 s. Find (a) the acceleration, (b) the average speed, and (c) the final speed.**

(a) (tool 5) $x = \dfrac{1}{2}at^2 + v_0 t$

$\dfrac{2x}{t^2} = a = \dfrac{2(245m)}{(34s)^2}$

$a = .424 \text{ m/s}^2$

(b) (tool 1) $x = v_{av} t \Rightarrow \dfrac{245m}{34s} = v_{av} = 7.21 \dfrac{m}{s}$

(c) (tool 3) $v = v_0 + at \Rightarrow v = \left(.424 \dfrac{m}{s^2}\right) 34s$

$v \approx 14.4 \text{ m/s}$

(tool 4) $v^2 = v_0^2 + 2ax \Rightarrow v^2 \approx 208 \dfrac{m^2}{s^2}$

$\Rightarrow v = \pm 14.4 \dfrac{m}{s} \Rightarrow v = 14.4 \dfrac{m}{s}$

❐❐ An object is moving to the right with a velocity of 14 m/s, and 11 s later it is moving to the left with a velocity of 19 m/s. Find (a) the acceleration, (b) the average velocity, (c) the maximum displacement of the object to the right during the 11 s interval, and (d) the position of the object at t = 11s.

(a) (tool 3) $v = v_0 + at \Rightarrow a = \dfrac{v - v_0}{t} = \dfrac{-19\frac{m}{s} - 14\frac{m}{s}}{11s}$

$$a = -3 \text{ m/s}^2 \text{ or } 3 \text{ m/s}^2 \text{ to the left}$$

(b) (tool 2) $v_{av} = \dfrac{v_0 + v}{2} = \dfrac{14\frac{m}{s} - 19\frac{m}{s}}{2} = -2.5\dfrac{m}{s}$

(c) (tool 4) $v^2 = v_0^2 + 2ax$

$$\dfrac{v^2 - v_0^2}{2a} = x_{max} = \dfrac{0^2 - \left(14\frac{m}{s}\right)^2}{2\left(-3\frac{m}{s^2}\right)} \approx 32.7 m$$

$v_0 = 14$ m/s $v = 0$

✗━━━━━━━━━━━┤

(d) (tool 5) $x = \dfrac{1}{2}at^2 + v_0 t = \dfrac{1}{2}\left(-3\dfrac{m}{s^2}\right)(11s)^2 + \left(14\dfrac{m}{s}\right)11s$

$$x = -27.5 m$$
$$t = 11 s$$

(tool 6) $x = v_{av} t$

$$= (-2.5 \text{ m/s})(11s) = -27.5 m$$

❐❐ A route driver has to make a 150 mile trip. For the first third of the trip her average speed is 45 mi/hr. What must be her average speed over the last 100 miles in order for her average speed for the whole trip to be 60 mi/hr?

whole trip: $x = v_{av} t \Rightarrow 150 \text{ miles} = (60 \text{mi/hr})t$
2.5 hr $= t =$ time for entire trip at an average speed of 60 mi/hr

first 50 miles $x = V_{AV} t \Rightarrow 50 \text{ miles} = (45 \text{ mi/hr})t$
$\Rightarrow t = 1.11$ hr

last 100 miles: (only 1.39 hours to go 100 miles)

$x = v_{av}t \Rightarrow 100 \text{ miles} = v_{av}(1.39)$

$v_{av} = 71.9 \text{ mi/hr}$

☐☐ In 1954 John Stapp rode a rocket propelled sled that was travelling at 632 mi/hr (282 m/s) that was stopped in 1.4 s. Find (a) the acceleration that acted in bringing the shed to halt, (b) over what distance the acceleration acted, and (c) how many times greater than the acceleration due to gravity the answer to (a) is.

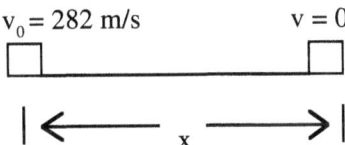

(a) $v = v_0 + at \Rightarrow a = \dfrac{v - v_0}{t} = \dfrac{0 - 282 \frac{m}{s}}{1.4s}$

$= -201 \text{ m/s}^2$

(b) (tool 1) $x = v_{av}t = \left(\dfrac{282\frac{m}{s} + 0}{2}\right)1.4s \cong 197m$

(tool 4) $v^2 = v_0^2 + 2ax \Rightarrow \dfrac{v^2 - v_0^2}{2a} = \dfrac{0^2 - (282 m/s)^2}{2(-201 m/s^2)} \cong 197m$

tool 5 could also be used

(c) $g = -9.8 \text{ m/s}^2$ thus $a = -201 \dfrac{m}{s^2}\left(\dfrac{1g}{-9.8 \frac{m}{s^2}}\right)$

and $a \cong 20.5g$

☐☐ **A bus starts from rest and accelerates at 2 m/s² for 12 s. It then travels at a constant speed for 20s, and then slows to a halt with an acceleration of -3.5 m/s². Find the total distance the bus traveled.**

The acceleration is not constant, thus the motion has to be broken up into parts where the acceleration is constant.

$x_T = x_1 + x_2 + x_3$

to find x_1: $x_1 = \dfrac{1}{2}a_1t_1^2 + (v_0)_1 t$

$\Rightarrow x_1 = \dfrac{1}{2}\left(2\dfrac{m}{s^2}\right)(12s)^2 = 144m$

to find x_2: for interval 1, $v = v_0 + at$

$v = 0 + 2 \text{ m/s}^2(12s) = 24 \text{ m/s}$

the velocity over the second interval is 24 m/s, thus,

$x_2 = (v_{av})_2 t_2 = \left(24 \dfrac{m}{s}\right)(20s) = 480m$

to find x_3: $v^2 = v_0^2 + 2a_3 x_3$

$$\dfrac{v^2 - v_0^2}{2a_3} = x_3 = \dfrac{0 - \left(24 \dfrac{m}{s}\right)^2}{2\left(-3.5 \dfrac{m}{s^2}\right)}$$

$x_3 \approx 82m$

Thus, $x_T + x_2 + x_3 = 706$ m

☐☐ A car and a truck start from rest at the same time, with the car some distance behind the truck. The truck has an acceleration of 2 m/s², and the car has an acceleration of 3m/s². The car is even with the truck after the truck has traveled 75 m.
(a) How long does it take the car to be even with the truck?
(b) How far behind the truck was the car initially?
(c) What is the velocity of the car and truck when they are even?

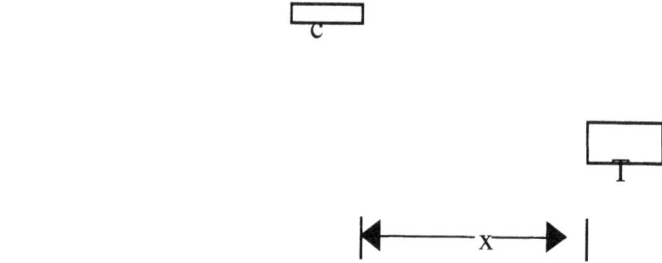

(a) $x_T = \dfrac{1}{2} a_T t^2 + (v_0)_T t$

$\Rightarrow 75m = \dfrac{1}{2}\left(2 \dfrac{m}{s^2}\right) t^2 \Rightarrow t = 8.66s$

(b) The car has traveled x + 75 m when it is even with the truck.

$x_c = \dfrac{1}{2} a_c t^2 + (v_0)_c t$

$x + 75m = \dfrac{1}{2}\left(3 \dfrac{m}{s^2}\right)(8.66s)^2$

$x_c = 112.5m - 75m = 37.5m$

(c) When the truck and the car are even,

$v_c = (v_0)_c + a_c t = 0 + (3m/s^2)(8.66s) \approx 26$ m/s

$v_T = (v_0)_T + a_T t = 0 + (2m/s^2)(8.66s) \approx 17.3$ m/s

☐☐ A ball is dropped from a roof and falls past the window below. It takes .25 s to pass the 3m height of the window. How far above the top of the window is the roof?

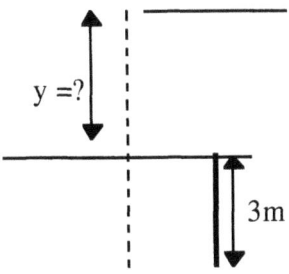

from the roof to the top of window:
$v_0 = 0$ and $a = -9.8$ m/s^2

for the window, $y = -3$m, $t = .25$ s
$a = -9.8$ m/s^2
$y = 1/2\, at^2 + v_0 t \Rightarrow -3\text{m} = 1/2\,(-9.8\text{ m/s}^2)(.25\text{s})^2 + (.25\text{s})v_0$
$\Rightarrow v_0 = -10.8$ m/s

Thus, at the top of window, the velocity of the ball is -10.8 m/s.
Using tool 4, $v^2 = v_0^2 + 2ay$
$(-10.8 \text{ m/s})^2 = 0^2 + 2(-9.8 \text{ m/s}^2)\, y$
$y = -5.95$ m

Thus, the roof is 5.95 m above the top of the window.

☐☐ A rocket lifts off with an acceleration of 5 m/s^2. At a height of 500m, the engine shuts down. How long is the rocket in the air? (Assume the rocket's motion is in the vertical sense.)

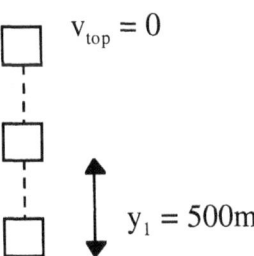

$t_{total} = t_1 + t_2 + t_3$

to find t_1: $\quad a_1 = 5$ m/s^2 ; $y_1 = 500$ m; $(v_0)_1 = 0$

$y = 1/2\, at^2 + v_0 t_1$

$500 = 1/2\,(5\text{m/s}^2)\, t_1^2 \Rightarrow t_1 = 14.1$ s to find t_2: $(v_0)_2 = v_1$

$$a = -9.8 \text{ m/s}^2$$

$$v_1 = (v_0)_1 + a_1 t = 0 + 5 \text{ m/s}^2 (14.1s) = 70.5 \text{ m/s}$$

Thus, $\quad v_2 = (v_0)_2 + a_2 t_2 \Rightarrow 0 = (70.5 \text{ m/s}) - 9.8 \text{ m/s}^2 (t_2)$

$$\Rightarrow t_2 = 7.19s$$

to find t_3: $\quad y_2 = (v_{av}) t_2 = \left(\dfrac{70.5 \frac{m}{s} + 0}{2} \right)(7.19s)$

$$y_2 = 253m$$

Thus, $\quad y_3 = -753$ m (the distance the rocket falls back down)
$\quad y_3 = 1/2\, a_3\, t_3^2 + (v_0)t_3$

$$-753 \text{ m} = 1/2\, (-9.8 \text{ m/s}^2)\, t_3^2 + 0 \Rightarrow t_3 = 12.4s$$

Thus, the rocket is in the air for 14.1s + 7.19s + 12.4s or
$\quad t_{total} = 33.7s$

☐☐ A person on a building 75 m high tries to drop a water-filled balloon on a professor walking toward the building as in the figure. The balloon is dropped when the professor is 5 m from the door. The balloon misses, striking the sidewalk 1m in front of the professor. What is the walking speed of the professor?

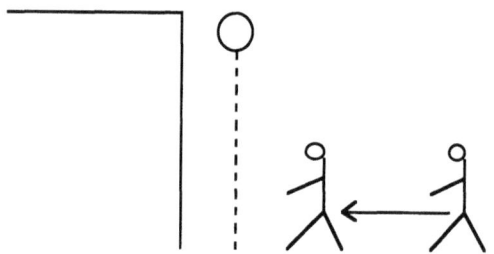

$v_0 = 0$, $a = -9.8 \text{m/s}^2$, y at impact $= -75$ m

The time it takes the balloon to fall 75 m is the time it takes the professor to walk 4 m

$$x_p = (v_{av})_p\, t \Rightarrow (v_{av})_p = \dfrac{x_p}{t}$$

for the balloon, $y = 1/2 at^2 + v_0 t$

$-75m = 1/2(-9.8 \text{ m/s}^2)\, t^2$

$t = 3.91$ s

Thus, $(v_{av})_p = 4m/3.91s = 1.02$ m/s

☐☐ At the instant a traffic light turns green, a car that has been waiting starts forward with an acceleration of 1.8 m/s². At the same moment, a truck travelling at a constant 9 m/s passes the car. (a) How long does it take the car to overtake the truck? (b) How far beyond the starting point will the car and truck be even? (c) How fast will the car be travelling when it is even with the truck?

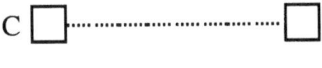

(a) $(v_0)_c = 0$
$a_c = 1.8$ m/s
$v_T = 9$ m/s = constant
when the car and thr truck are even,
$x_c = x_T \Rightarrow \frac{1}{2} a_c t^2 = (v_{av})_T t$

$\Rightarrow \frac{1}{2}\left(1.8 \frac{m}{s^2}\right) t^2 = \left(9 \frac{m}{s}\right) t$

$t = 10$ s

(b) $x_T = x_c$ when even $\Rightarrow (v_{av})_T t = x_T$

$90m = x_T = x_c$

(c) $v_c = (v_0)_c + a_c t$
$= 0 + (1.8$ m/s²$) 10$s
$= 18$ m/s

☐☐ A ball is thrown with an initial velocity of 11 m/s down as it leaves the thrower's hand.
(a) What is the velocity of the ball when 3 seconds have elapsed?
(b) What is the velocity of the ball when it has traveled 9m down?
(c) What acceleration acted on the ball if it was in the thrower's hand over a distance of 8m?
(d) What acceleration acts on the ball after it leaves the thrower's hand?
(e) If the ball initially is 35m above the ground, how long does it take to hit the ground?

(a) $v = v_0 + at$
$= -11$m/s -9.8m/s²$(3$s$)$
$= -40.4$ m/s

(b) $v^2 = v_0^2 + 2ay$
$v^2 = (11$m/s$)^2 + 2(-9.8$m/s²$)(-9$m$) \approx 297$ m²/s² $\Rightarrow v \approx -17.2$ m/s

(c) $\frac{v^2 - v_0^2}{2y} = a = \frac{\left(-11 \frac{m}{s}\right)^2 - 0^2}{2(-.8m)} \cong -75.6 \frac{m}{s^2}$

(d) $a = g = -9.8$ m/s²

(e) $x = 1/2 \, at^2 + v_0 t$

$-35m = 1/2(-9.8$ m/s²$) t^2 - (11$ m/s$) t \Rightarrow t = -4.02$s or $t = 1.78$s
Thus, it takes 1.78s to hit the ground.

TWO-DIMENSIONAL MOTION AND RELATIVE VELOCITY

THE FUNDAMENTAL IDEA HERE IS:

That motion in x is independent of motion in y. This means that to work a two-dimensional motion problem, each part should be worked separately, using the five tools of motion that are used in solving one-dimensional motion problems. These five tools are outlined in Chapter II. Some basic assumptions and relationships are as follows:

- We assume that the air does not resist the motion of an object through it, i.e., $a_x = 0 \Rightarrow v_x$ = constant.

- Tool 5 states that $y = 1/2 at^2 + v_0 t$, while tool 1 says that $x = (v_{av})_x \, t$. By putting these together, $y = y(x) = (\tan\theta)x - \dfrac{gx^2}{2v_0^2 \cos^2\theta}$ which shows that projectile motion is parabolic in the absence of air resistance to the motion. [Note that the general form of a parabola is $y = a + bx + cx^2$].

- For an object executing uniform (constant speed) circular motion, the acceleration that acts is called radial acceleration (or centripetal acceleration) and is given by v^2/r.

- Relative velocity is velocity relative to a particular reference frame. A formal development leads to the Galilean velocity transform, $\vec{v}_1 = \vec{v}_2 + \vec{v}_{21}$ where,

  $\vec{v}_1$ = velocity in the at rest reference frame

  $\vec{v}_2$ = velocity in the moving reference frame

  $\vec{v}_{21}$ = velocity of moving reference frame relative to the at rest reference frame

- A less formal development says that relative velocity is the resultant of two or more velocity vectors acting on an object.

SOLVED PROBLEMS

□□ An object is given a horizontal velocity of 5 m/s from the roof of a building that is 18 m high. Find (a) the time of flight, (b) the x displacement at impact, (c) the components of velocity at impact, and (d) the magnitude and the direction of the velocity vector at impact.

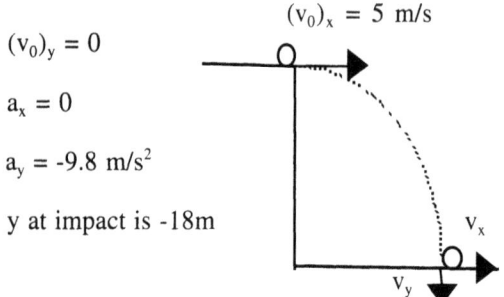

$(v_0)_y = 0$

$a_x = 0$

$a_y = -9.8$ m/s²

y at impact is -18m

(a) $y = 1/2\, a_y t^2$

$-18m = 1/2(-9.8m/s^2)t^2$

$\dfrac{-36m}{-9.8\,\frac{m}{s^2}} = t^2 \Rightarrow t \cong 1.92s$

(b) $x = (v_{av})_x t$

$= (5m/s)1.92s$

$x_{impact} = 9.6m$

(c) $v_x = (v_0)_x = 5$m/s because of the assumption that $a_x = 0$

$v_y = (v_0)_y + a_y t \Rightarrow v_y = -18.8$ m/s

(d) $v = \sqrt{v_x^2 + v_y^2} \cong 19.5\,\dfrac{m}{s}$

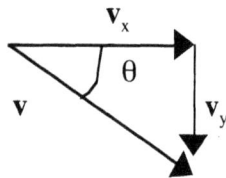

$\tan\theta = \dfrac{v_y}{v_x} = \dfrac{18.8\,\frac{m}{s}}{5\,\frac{m}{s}} = 3.6 \Rightarrow \theta = 75.1°$

Thus, $\vec{v} \cong 19.5\,\dfrac{m}{s}$ @ 75.1° below horizontal at impact.

Note that rather than using the components of vectors to determine the direction of the vector, by using a sketch and the magnitudes of the components of the vector one can get the direction without having to remember that the tangent is positive in the first and third quadrants, and negative in the second and fourth quadrants.

☐☐ For the flat earth, frictionless atmosphere problem, find an expression in terms of the initial speed v_0, the launch angle θ, and the acceleration due to gravity g for (a) time of flight, (b) the horizontal range (maximum displacement in x), (c) the maximum height to which the projectile rises, and (d) that angle θ of launch for which x is a maximum.

(a) to find the time of flight, note that the time of rise = the time of fall and therefore, time$_{rise}$ = $2t_{flight}$.

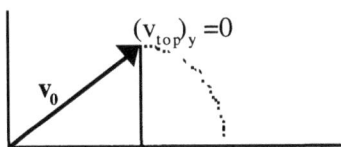

Looking at the y component of the motion,

$v_y = (v_0)_y + a_y t \Rightarrow 0 = v_0 \sin\theta - gt_{rise}$

$\Rightarrow t_{rise} = \dfrac{v_0 \sin\theta}{g}$ so that $t_{flight} = 2t_{rise} = \dfrac{2v_0 \sin\theta}{g}$

(b) $x_{max} = (v_{av})_x \, t_{flight} = R$

$R = (v_0 \cos\theta)\dfrac{2v_0 \sin\theta}{g} = \dfrac{v_0^2 \sin 2\theta}{g}$

because $2\cos(x)\sin(x) = \sin(2x)$

(c) $y_{max} = H = (v_{av})_y \, t_{rise} = \left(\dfrac{v_0 \sin\theta + 0}{2}\right)\left(\dfrac{v_0 \sin\theta}{g}\right)$

$H = \dfrac{v_0^2 \sin^2\theta}{2g}$

(d) Since $R = x_{max} = \dfrac{v_0^2 \sin 2\theta}{g}$, R is a maximum if $\sin 2\theta = 1$

$\Rightarrow \sin(\pi/2) = 1$

$\Rightarrow 2\theta = \pi/2$ so that $\theta = \pi/4$ radians or $\theta = 45°$.

☐☐ The earth is 1.49×10^{11} m from the sun and takes 365 days to complete one orbit around the sun. What radial acceleration acts on the earth because it is orbiting the sun?

$a_{radial} = v^2/r$ and displacement $= v_{av} t \Rightarrow \dfrac{2\pi r}{t} = v_{av}$

$\Rightarrow \dfrac{(6.28)(1.49 \times 10^{11})}{(365)(86,400 s)} \approx 29,700 \dfrac{m}{s} = v_{av}$

so that $a_{radial} = \dfrac{\left(29{,}700\,\dfrac{m}{s}\right)^2}{1.49 \times 10^{11}\,m}$

$= 5.91 \times 10^{-3}\,m/s^2$

☐☐ **A hot air balloon is rising at 6 m/s. An object is dropped when the balloon is 25m above the ground. Find (a) the initial velocity of the object relative to the balloon, (b) the initial velocity of the balloon relative to the ground, and (c) the time it takes for the object to hit the ground.**

(a) relative to the balloon the initial velocity is 0.

(b) relative to the ground, the initial velocity is 6 m/s

(c) (working in the ground reference frame): $y = 1/2\,at^2 + v_0 t \Rightarrow$

$-25m = \dfrac{1}{2}\left(-9.8\,\dfrac{m}{s^2}\right)t^2 + \left(6\,\dfrac{m}{s}\right)t$

or $0 = -4.9t^2 + 6t + 25$
$t = 2.95s$ or $-1.72s$.

Thus, it takes 2.95s for the object to hit ground.

OR

(c) (working in the reference frame of the balloon): $x_{balloon} = -x_{obs}$ at impact

$(6m/s)t + 25\,m = 1/2(at^2) \Rightarrow 6t + 25 = +4.9t^2$
$\Rightarrow 0 = +4.9t^2 - 6t - 25$
$\Rightarrow t = 2.95s$ or $-1.72s$

So it takes 2.95s to hit the ground.

☐☐ **A ball is thrown with a velocity of 15m/s at an angle of 30° relative to the horizontal. A catcher is standing 30 m from where the ball is thrown. (a) Will the catcher have to run toward the thrower or away from the thrower to catch the ball? (b) What must be the average velocity of the catcher to catch the ball?**

(a) $t_{flight} = \dfrac{2v_0 \sin\theta}{g} = \dfrac{20\,\dfrac{m}{s}}{9.8\,\dfrac{m}{s^2}} = 1.53s$

Range of the ball $= (v_0 \cos\theta)t_f \cong 13\,\dfrac{m}{s}(1.53s) = 19.9m$

so catcher runs toward the thrower.

(b) $x = v_{av}t \Rightarrow v_{av} = \dfrac{x}{t} = \dfrac{30m - 19.9m}{1.53s} \cong 6.6\,\dfrac{m}{s}$

The average velocity of catcher must be about 6.6 m/s toward catcher in order to catch the ball.

☐☐ An object is thrown at an angle of 37° above the horizontal and returns to the same level after travelling horizontally 25 m. (a) Find the initial velocity of the object, and (b) find the time of flight of the object.

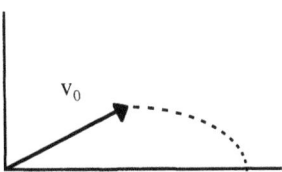

(a) $y = x\tan\theta - \dfrac{gx^2}{2v_0^2 \cos^2\theta}$

at impact $y = 0 \Rightarrow x\tan\theta = \dfrac{gx^2}{2v_0^2 \cos^2\theta}$

$\Rightarrow v_0^2 = \dfrac{gx}{2\tan\theta \cos^2\theta} = \dfrac{\left(9.8\,\dfrac{m}{s^2}\right)25m}{2(\tan 37°)(\cos 37°)^2}$

$v_0^2 \approx 255\,\dfrac{m^2}{s^2} \Rightarrow v_0 \cong 16\,\dfrac{m}{s}$

(b) $x = (v_{av})_x t$
$(v_{av})_x = v_0\cos 37° = 12.8$ m/s
$25m = (12.8 m/s)t \Rightarrow t_{flight} = 1.95$ s

☐☐ A bullet is fired from a rifle that is 1.7 m above the ground, with a velocity of 338 m/s. If the rifle is horizontal, how long does it take the bullet to hit the ground?

$(v_0)_y = 0$

$(v_0)_x = 338$ m/s

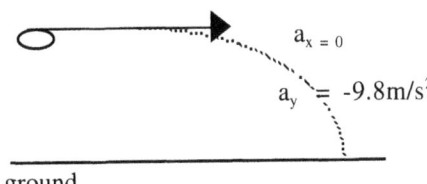

ground
at impact, y = -1.6m

since motion in x does not affect motion in y, $[y = 1/2\, a_y t^2 + (v_0)_y t]$

$-1.6m = 1/2(-9.8\ m/s^2)t^2$

$\Rightarrow t = 0.57$ s to hit the ground

☐☐ A person is running along a roof, as in the figure, when she jumps to the next building which is 3 m away. The second building is 2.5 m below the first. What horizontal velocity must she have to just make the edge of the second building?

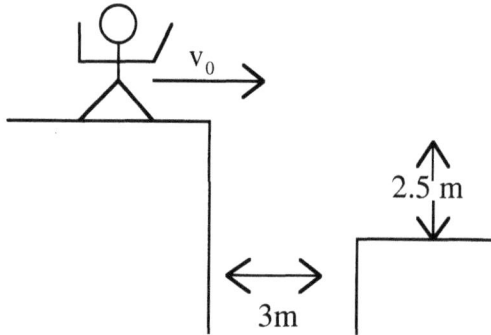

Here $a_y = -9.8$ m/s², $a_x = 0$, $(v_0)_y = 0$.
If she just makes the edge of the second building, she will have displaced in the vertical sense -2.5m.
Thus, $y = 1/2\, a_y\, t^2 + (v_0)_y\, t$

-2.5 m $= 1/2\,(-9.8$ m/s²$)\, t^2$
$t = .714$s

$x = (v_{av})_t t \Rightarrow \dfrac{x}{t} = (v_{av})_x = \dfrac{3m}{.714s} = 4.2\,\dfrac{m}{s}$

$x = (v_{av})(t)$

Her initial horizontal velocity must be at least 4.2 m/s.

☐☐ In football, "hang time" is how long the football stays in the air after a punt. If the ball stays in the air 4.6 s, and has a horizontal displacement of 45 yards (41.2 m), find (a) the angle at which the ball is kicked, and (b) the initial speed of the ball. (Assume the ball returns to the same level at which it was kicked.)

(a) at impact $y = 0$ $a_x = 0$
$\qquad a_y = -9.8$ m/s
$\qquad y = 1/2\, a_y t^2 + (v_0)_y t \quad t = 4.6$ s
$\qquad 0 = 1/2\,(-9.8$ m/s²$)(4.6s)^2 + (4.6s)(v_0)_y$
$\qquad (v_0)_y = 22.5$ m/s
$\qquad x = (v_{av})_x t \Rightarrow (41.2m) = (v_{av})_x (4.6s)$
$\Rightarrow (v_{av})_x = 8.96$ m/s $= (v_0)_x$

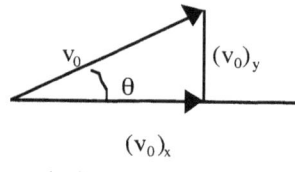

$\tan \theta = \dfrac{(v_0)_y}{(v_0)_x}$

$\tan \theta = 2.51$
$\theta = 68.3°$

(b) $v = \sqrt{(v_0)_x^2 + (v_0)_y^2} = \sqrt{\left(8.46\frac{m}{s}\right)^2 + \left(22.5\frac{m}{s}\right)^2}$

v = 24.2 m/s

☐☐ A pilot wishes to fly west on a day when there is a south wind of 80 km/hr. The airspeed [speed of plane with respect to the air] is 240 km/hr. (a) In what direction should the pilot head in order to fly west? (b) What is the speed of the plane relative to the ground?

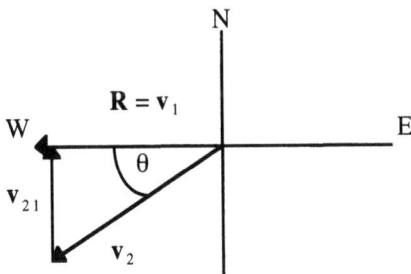

(a) $\tan\theta = \frac{v_{21}}{v_2} = \frac{80 km/hr}{240 km/hr} \Rightarrow \theta = 18.4°$

Thus, pilot should head 18.4° S of W.

(b) $R^2 + v_{21}^2 = v_2^2$

$R = \sqrt{v_2^2 - v_{21}^2} \Rightarrow R = 226 km/hr$

☐☐ A helicopter rotor has four blades each of length 3m, and the rotor is rotating at 1500 rev/min. (a) what is the radial acceleration acting on the tip of a blade, and (b) how many g's is this acceleration?

(a) 1 rev = $2\pi r$ = (6.28)3m = 18.8m

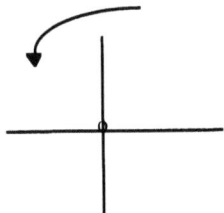

$a_{radial} = v^2/r$ $v = 1500\left(\frac{rev}{min}\right)\left(\frac{1 min}{60 s}\right)\left(\frac{18.8 m}{1 rev}\right)$

v = 471 m/s

Thus, $a_{radial} = \frac{v^2}{r} = \frac{(471 m/s)^2}{3m} = 73950 m/s^2$

(b) $a_{radial} = 73950 m/s^2 \left(\frac{1g}{9.8 m/s^2}\right) = 7.55 \times 10^3 g$

❏❏ A river flows east with a velocity of 2.5 m/s. A girl rows north with a velocity of 3 m/s relative to the water. If the river is 600 m wide, (a) how long does it take to cross the river, (b) how far downstream does the girl land on the opposite bank, and (c) if the girl wanted to reach the opposite bank at a point directly opposite from where she started, in what direction should she head?

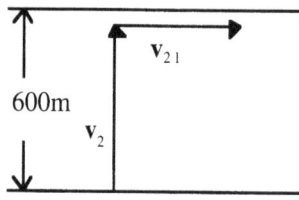

(a) $x_N = (v_{av})_N t$

  $600m = (3m/s)t$

  $t = 200s$

(b) $x_E = (v_{av})_E t = (2.5m/s)(200s) = 500 m$

(c) If the Resultant velocity is to directly opposite the starting point,

  $\sin\theta = \dfrac{v_{21}}{v_2} = \dfrac{2.5 m/s}{3 m/s} = .833$

  $\theta = 56.4°$

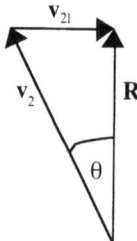

Thus, she should head 56.4° W of N or 56.4° upstream.

❏❏ The driver of a car that is behind a truck wishes to pass the truck which is travelling at 18 m/s. The car is also at an initial speed of 18 m/s. The car's maximum acceleration is 0.5 m/s^2, and the car is initially 25m behind the truck. The car is 5m long and the truck is 20 m long. The car pulls back into the truck's lane when the car is 25m in front of the truck. Using the reference frame of the truck, find (a) how long it takes to pass the truck, and (b) what is the final speed of the car relative to the truck?

(a) In the reference frame of the truck, v_0 of the car is 0, and the distance relative to the truck that the car has to travel is 75m.

$$x = \frac{1}{2}at^2 + v_0 t$$

$$75m = \frac{1}{2}\left(.5\frac{m}{s^2}\right)t^2$$

$t = 17.3s$
to pass the truck.

(b) Relative to the truck, the final speed of the car is $v = v_0 + at$
$= 0 + (.5m/s^2)(17.3s)$
$v = 8.65$ m/s

□□ For the previous question, using the earth as a reference frame, find (a) How long it takes the car to pass the truck, (b) how fast the car is travelling when it returns to the truck's lane, and (c) how far the car travels while it is passing the truck.

(a) $x_T = x_C - 75m = (1/2)a_c t^2 + (v_0)_c t - 75m$

$(18m/s)t = 1/2(.5m/s^2)t^2 + (18m/s)t - 75m$

so that $75m = 1/2(.5m/s^2)t^2 \Rightarrow t = 17.3s$ as before.

(b) $v = v_0 + at$, so $v = 18$ m/s $+ (.5m/s^2)17.3s$
$v = 26.7s$

(c) Since $x_T = x_C - 75m$,

$x_C = x_T + 75m = (v_T)t + 75m$

$x_C = 386m$

□□ A plane is diving at an angle of 37° and drops a flare at an altitude (height) of 500m. The flare hits the ground 5 s after its release. Find (a) the speed of the plane, (b) how far the flare traveled horizontally before impact and (c) the speed of the flare at impact.

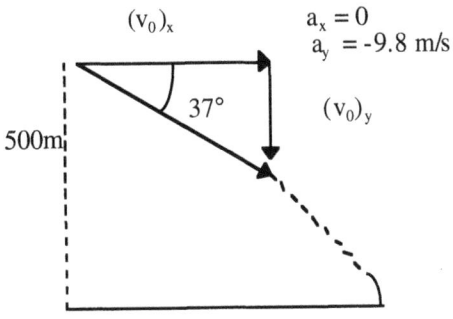

(a) $y = 1/2(a_y)t^2 + (v_0)_y t$

$-500m = 1/2(-9.8m/s^2)(5s)^2 + (v_0)_y(5s)$

$-75.5m/s = (v_0)_y$

$$\tan 37° = \frac{(v_0)_y}{(v_0)_x}$$

$$(v_0)_x = \frac{75.5 m/s}{\tan 37°} = 100 m/s$$

$$v_0 = (v_0)_x + (v_0)_y = \sqrt{(100m/s)^2 + (-75.5m/s)^2} = 125 m/s$$

(b) $x = (v_{av})_x t = (100m/s)5 = 500m$

(c) at impact, $v_x = 100$ m/s
$v_y = (v_0)_y + a_y t$

$\qquad = -75.5$m/s $+ (-9.8$m/s$^2)$t

$v_y = -125$m/s

$$v = \sqrt{v_x^2 + v_y^2} = \sqrt{(100m/s)^2 + (-125m/s)^2} \approx 160 m/s$$

☐☐ **A small boat on a lake moves at 20 knots west and the wind is blowing south at 15 knots. In which direction will a flag point to an observer on the boat?**

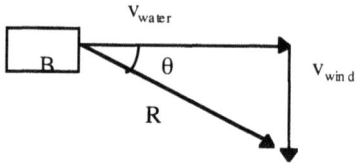

On the boat, the observer sees the water moving at 20 knots, and the wind is blowing south at 15 knots. Thus,

$$\tan \theta = \frac{v_{wind}}{v_{water}} = \frac{15 knots}{20 knots} = 0.75$$
$$\Rightarrow \theta = 36.9°$$

The observer sees the flag flying at 36.9° S of W.

☐☐ **A child throws a ball on the roof of an adjacent building as in the figure. If the ball lands on the roof 3.8s after being thrown, find (a) $(v_0)_x$, (b) $(v_0)_y$, and (c) the angle at which the ball was launched.**

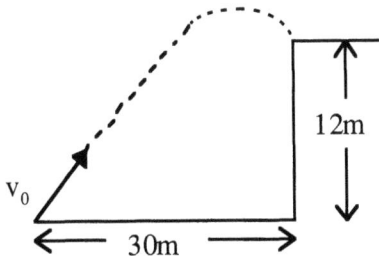

(a) $x = (v_{av})_x t \Rightarrow 30m = (v_{av})_x (3.8s) \Rightarrow (v_0)_x = 7.89 m/s$

(b) $y = \frac{1}{2} a_y t^2 + (v_0)_y t \Rightarrow 12m = \frac{1}{2}(-9.8 m/s^2)(3.8s)^2 + (3.8)(v_0)_y$

$(v_0)_y = 21.8$ m/s

(c)

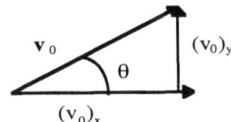

$\tan \theta = \dfrac{(v_0)_y}{(v_0)_x} = \dfrac{21.8 m/s}{7.89 m/s} = 2.76$

$\theta = 70°$

The ball was thrown at 70° above the horizontal.

☐☐ **How fast in rev/min would a centrifuge have to turn to cause an acceleration of 500 g (500 times larger than the acceleration due to gravity) on a specimen that is 4 cm from the center of rotation?**

$a_{radial} = v^2/r \Rightarrow v^2 = (a_{radial}) r$

$\qquad = (500)(9.8 m/s^2).04m$

$\qquad v^2 = 196 \; m^2/s^2 \Rightarrow v = 14 m/s$

1 rev $= 2\pi r = 6.28(.04m) = .251m$ and 1 min = 60 s

$v = 14 \dfrac{m}{s} \left(\dfrac{60s}{1 min}\right)\left(\dfrac{1 rev}{.251m}\right) \approx 3340 \dfrac{rev}{min}$

☐☐ A ball is thrown up with a velocity of 30 m/s at an angle of 35° with respect to the horizontal. A 10m high fence is 40m from the point where the ball is thrown. Will the ball fall in front of the fence, hit the fence, or go over the fence?

To find y when x = 40m, $y = (\tan\theta)x - \dfrac{gx^2}{2v_0^2 \cos^2\theta} = (40m)\tan 35° - \dfrac{9.8 m/s^2 (45m)^2}{2(30m/s)^2(\cos 35°)^2}$

y = 28m − 16.4m

so $y_{x=30m}$ = 11.6m and the ball will go over the fence.

☐☐ Find the radial acceleration of an object at the equator of the earth (radius of earth is 6400km).

$a_{radial} = v^2/r$ the velocity of the earth is $\dfrac{1 rev}{day}\left(\dfrac{1 day}{86400 s}\right)\left(\dfrac{2\pi[6.4 \times 10^6 m]}{1 rev}\right) = 465 m/s$ at the equator

$a_{radial} = \dfrac{(465 m/s)^2}{6.4 \times 10^6 m} = 3.38 \times 10^{-2} m/s^2$

☐☐ For the previous problem, if the air were perfectly still, what would be the speed of the wind in miles/hr, relative to an observer at the equator of the earth?

Since the speed of a point on the equator is 465m/s, relative to an observer at the equator the wind would appear to be moving at $465 m/s \left(\dfrac{1 mile}{1609 m}\right)\left(\dfrac{3600 s}{1 hr}\right)$ or 1040 mi/hr.

☐☐ A boat moves east at 25 km/hr. A wind is blowing from the northwest at 10 km/hr. In what direction would a flag on the mast of the boat point as seen by an observer on the boat?

vector	W	S
V_{water}	25km/hr	
V_{wind}	-7.07km/hr	7.07km/hr
R	17.9km/hr	7.07km/hr

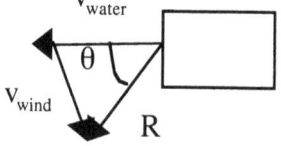

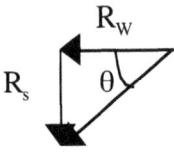

$$\tan\theta = \frac{R_s}{R_w} = \frac{7.07 km/m}{17.9 km/m} = .395 \Rightarrow \theta = 21.5°$$

So the flag points 21.5° S of W

NEWTON'S LAWS AND FRICTION

THE FUNDAMENTAL IDEAS HERE ARE NEWTON'S THREE LAWS AND THEIR APPLICATIONS:

- A force is "push" or "pull" that acts on an object.

- Mass is quantity of matter.

- Newton's 1st Law: Every object persists in its state of motion unless acted upon by some external force (An object at rest remains at rest, an object in motion remains in motion unless acted upon by some external force).

- Newton's 2nd Law: $F_{unbalanced} = ma = F_{go} - F_{opposition}$

- Newton's 3rd Law: For every force there is an equal but opposite reaction.

- Friction is a force that opposes motion.

- The magnitude of the static (at rest) frictional force is given by $f_s \leq \mu_s N$, where μ_s is the coefficient of static friction and N is the magnitude of the normal force (perpendicular force with which the surface pushes back on the object).

- The static frictional force is equal to $\mu_s N$ if and only if the object is on the verge of sliding.

- The magnitude of the kinetic (moving) frictional force is given by $f_k = \mu_k N$, where μ_k = coefficient of kinetic friction.

- From Newton's 1st Law, one has the following ideas:
  (a) inertia - that property of matter whereby it resists any change in its state of motion

  (b) mass is a measure of inertia

  (c) equilibrium - an object at rest or moving with constant velocity

  (d) If the object is in a condition of equilibrium, then the vector sum of all forces acting on it are 0, i.e. $\Sigma \vec{F} = 0$. Thus, $\Sigma F_x = 0$ and $\Sigma F_y = 0$ and $\Sigma F_z = 0$

  (e) Another way of saying this is if the object is in a condition of equilibrium,

  $(1) \Sigma F_{up} = \Sigma F_{down}$
  $(2) \Sigma F_{left} = \Sigma F_{right}$
  These are called the force conditions of equilibrium.

- From Newton's 2nd Law, one has the following ideas:

  (a) $\vec{F}_{un} = m\vec{a}$ is a vector equation

(b) $F_{un} = ma = F_{go} - F_{opp}$ means that F_{go} is the magnitude of the sum of the forces making an object move, and F_{opp} is the magnitude of the sum of the forces that oppose the object's motion

(c) weight is the force that "pulls" an object toward the center of the earth, $w = mg$

- To apply Newton's Laws as a method to solve mechanics problems, one isolates the object in question and then applies Newton's Laws. If the object is in a condition of equilibrium, then the force conditions of equilibrium apply. If there is acceleration, then Newton's 2nd Law applies.

- If an object is on a hill or incline, then for motion along the incline $w_{||} = w\sin\theta$ and $w_\perp = w\cos\theta$ as in the figure.

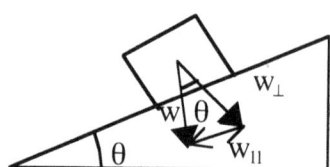

- A tension force is a force transmitted by a string or cable.

SOLVED PROBLEMS

❏❏ **What is the magnitude of the weight of a 45 kg object?**

$w = mg = (4.5\text{kg})(9.8\text{m/s}^2) = 44.1\text{N}$

❏❏ **For the adjacent figure, an object at rest is suspended by a cord. What is the tension in the cord?**

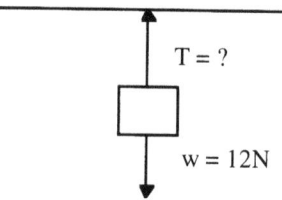

If the object is at rest, the force conditions of equilibrium apply. Thus, $\Sigma F_{up} = \Sigma F_{down} \Rightarrow T = w = 12N$

❏❏ **If the object in the previous question is pulled to the right with a horizontal force F so that the cord makes an angle of 34° with the vertical, find F and T.**

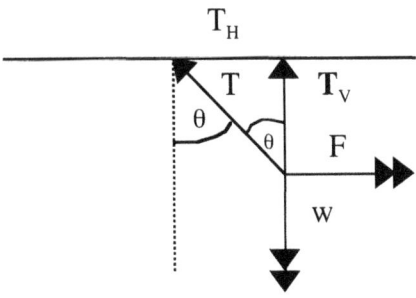

$\Sigma F_{up} = \Sigma F_{down}$
$\Sigma F_{left} = \Sigma F_{Right}$

Break T into components, T_V and T_H, $\quad T_V = T\cos\theta$
$\quad T_H = T\sin\theta$

Then $T_H = F$ and $T_V = w \quad \Rightarrow \dfrac{T\sin\theta}{T\cos\theta} = \dfrac{F}{w} \Rightarrow F = w\tan\theta = 29.7N$

$T = \dfrac{F}{\sin\theta} = 53.2N$

☐☐ A 6kg object is at rest on a horizontal surface as in the figure. A horizontal force F is applied to the box. If $\mu_K = .4$ and $\mu_S = .6$, find the frictional force if (a) F = 0, (b) F = 30N, and (c) F = 46N

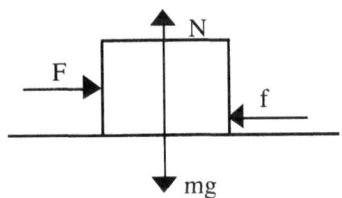

(a) if F = 0, this implies f = 0 because $\Sigma F_{left} = \Sigma F_{right}$

(b) Here it is necessary to find f_k and the maximum value of f_s.

$\Sigma F_{up} = \Sigma F_{down} \Rightarrow N = mg = 58.8N$,
Thus, $(f_s)_{max} = \mu_s N = (.6)(58.8N) = 35.3N$
$(f_k) = \mu_k N = 23.5N$

Thus, it takes 35.3 N for the box to slide, so if F = 30N $\Rightarrow f_s = 30N$

(c) if F = 46N, then $f_k = 23.5N$ because the maximum static frictional force is 35.3N. Thus, the box will begin to move. If the box is moving the frictional force will be kinetic.

☐☐ For the previous question, what would be the acceleration that acts on the box if F = 46N?

$F_{un} = ma = F_{go} - F_{opp} \Rightarrow (6kg)a = 46N - 23.5N$
$\qquad a = 3.75 \text{ m/s}^2$

☐☐ For the given figure, what is the magnitude of the normal force? (Assume block is at rest initially)

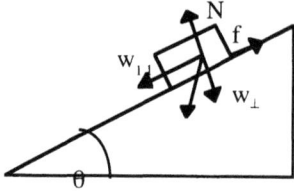

Here

$\Sigma F_{up} = \Sigma F_{down}$
$N = w_\perp = w\cos\theta$
$N = [(6kg)(9.8m/\sec^2)](\cos 39°)$ $\theta = 39°$ and $m = 6kg$
$N = (58.8N)\cos 39°$
$N = 45.7N$

☐☐ For the previous problem, if $\mu_s = .9$ and $\mu_k = .7$, (a) will the block slide down the incline? And (b) if so, what is its acceleration?

(a) The force down the incline (to the left) is given by $w_{||} = w\sin\theta$
$= (58.8N)\sin 39° = 37N$

The frictional force maximum is $f_s = \mu_s N = (.9)45.7N$
$f_s = 41.1N$
Thus, the force to the left (down incline) is 37N, and the maximum force to the right (up incline) is 41.1N. Hence, the block won't slide.

(b) a = 0, because the block is not moving.

☐☐ An object that weighs 800N is suspended by a cable as in the figure. Find the tension in the cable if (a) the acceleration is 0, (b) the acceleration is up at 8 m/s², and (c) the acceleration is down at 6 m/s².

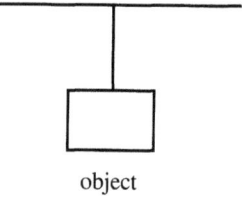

object

If there is acceleration, use Newton's 2nd Law. If the object is at rest or moving with constant velocity, use the force conditions of equilibrium.

(a) if $a = 0 \Rightarrow$ speed = constant and thus, $\Sigma F_{up} = \Sigma F_{down} \Rightarrow T = W = 800N$

(b) If a = 8 m/s² up, then $F_{un} = ma = F_{go} - F_{opp} \Rightarrow ma = T - W$
$(81.6kg)8m/s^2 = T - 800N$
$\Rightarrow T \approx 1450N$

(c) If a = 6m/s² down, $F_{un} = ma = F_{go} - F_{opp} \Rightarrow ma = W - T$

$(81.6\text{kg})6\text{m/s}^2 = 800\text{N} - T$
$\Rightarrow T \cong 310\text{N}$

Note that F_{go} is in the direction of the acceleration, while F_{opp} is the force or forces that oppose the motion.

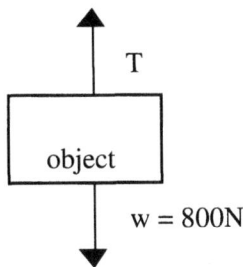

❏❏ **If an object is on an inclined plane, under what conditions will the frictional force be directed down the plane?**

If the object is moving up the plane.

❏❏ **A car has an emergency stop where all four wheels lock up. If the initial velocity of the car was 50mph (22.3m/s), and the coefficient of kinetic friction between the tires and the pavement is 0.8, find the length of the skid marks.**

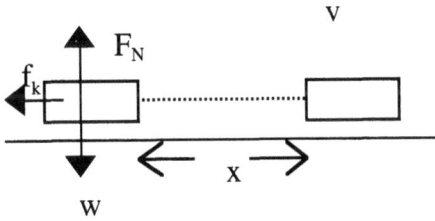

[Since car is not moving in vertical sense, $F_N = w = mg$]
$F_{un} = ma = F_{go} - F_{opp}$
$ma = 0 - f_k$
$ma = 0 - \mu_k F_N = -\mu_k mg \Rightarrow a = -\mu_k g = -7.84 \text{ m/s}^2$

Thus, $v^2 = v_0^2 + 2ax$

$$\Rightarrow \frac{v^2 - v_0^2}{2a} = X = \frac{0 - (22.3 m/s)^2}{2(-7.84 m/s^2)} = 31.7m$$

❏❏ **A person who weighs 850N is on a scale in an elevator. What does the scale read if the elevator is (a) at rest, (b) moving up with an acceleration of 12 m/s^2, and (c) moving down with an acceleration of 8 m/s^2? [mass = w/g = 86.7kg]**

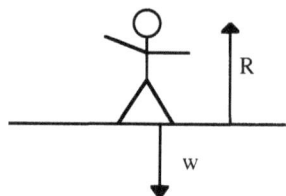

(a) If the elevator is at rest, $\Sigma F_{up} = \Sigma F_{down}$
$$R = w = 800N$$

(b) If elevator has an acceleration of 12 m/s² up, then Newton's 2nd Law says that
$F_{un} = ma = F_{go} - F_{opp}$ so that $(86.7kg)12m/s^2 = R - w$
$\Rightarrow 1040N = R - 850N \Rightarrow R = 1890N$

(c) If elevator has an acceleration of 8 m/s² down, then
$F_{un} = ma = F_{up} - F_{opp} \Rightarrow (86.7kg)(8m/s^2) = 850 - R$
$$R = 850N - 694N$$
$$R = 156N$$

□□ A girl pulls a 40 kg sled, as in the figure, at constant velocity. If the coefficient of kinetic friction is .3, what is the magnitude of force F?

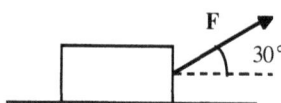

Since the velocity of the sled is constant, that means the force conditions of equilibrium apply.
Thus, $\Sigma F_{up} = \Sigma F_{down}$ $\Sigma F_{left} = \Sigma F_{rt}$

$N + F_V = w$ $f_k = F_H$

Thus, $N = w - F_V$ $u_k N = F\cos 30°$
$\quad = mg - F\sin 30°$

Thus, $u_k(mg - F\sin 30°) = F\cos 30°$

$\Rightarrow u_k mg = F\cos 30° + u_k F\sin 30°$
so that

$$\frac{u_k mg}{\cos 30° + u_k \sin 30°} = F = \frac{(.3)(40kg)(9.8m/s^2)}{.866 + (.3)(.5)} \Rightarrow F \cong 116N$$

□□ A rope can sustain a force of 750N. What is the minimum acceleration that a 900N person can have in sliding down the rope without breaking it?

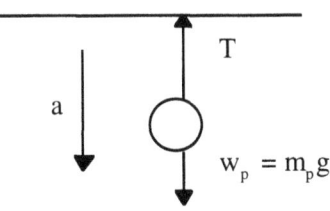

$w = mg \Rightarrow m = 91.8\text{kg}$

If $a = 0$, $T = 900\text{N}$ and rope breaks.

Thus, from Newton's 2nd Law $F_{un} = ma = F_{up} - F_{opp}$

$(91.8\text{kg})a = 900\text{N} - 750\text{N}$

$$a_{min} = \frac{150N}{91.8kg} = 1.63 m/s^2$$

☐☐ A picture is hung as in the figure. What is angle θ so that the tension T in each wire is equal to the weight?

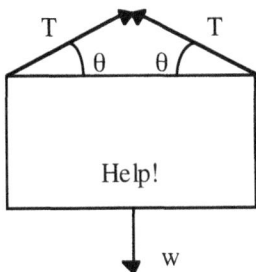

$\Sigma F_{up} = \Sigma F_{down}$
$T\sin\theta + T\sin\theta = w$, and since $T = w$,
$\Rightarrow 2\sin\theta = 1$
$\sin\theta = 1/2 \Rightarrow \theta = 30°$

☐☐ A steam catapult for launching aircraft from rest produces an acceleration of 3 g's. If an aircraft (mass 11,400 kg) is launched with a speed of 54 m/s, (a) what force is required, and (b) over what distance does the force act, and (c) what is the time the force acts on the aircraft?

(a) $F_{un} = ma = (11,400\text{kg})(3)(9.8\text{m/s}^2)$
    $F_{un} = 335,000\text{N}$

(b) $v^2 = v_0^2 + 2ax$
    $(54\text{m/s})^2 = 0 + 2(3)(9.8\text{m/s2})x \Rightarrow x = 49.6\text{m}$

(c) $x = v_{av}t \Rightarrow 49.6\text{m} = [27\text{m/s}]t$
    $t = 1.84\text{s}$

☐☐ A toy train consists of an engine (mass = 1.5kg) and 4 cars (mass of each car is .5 kg). The engine gives the train an acceleration of 0.8 m/s² by pulling on the first car with a force of 5N. Find the total frictional force that acts on the four cars.

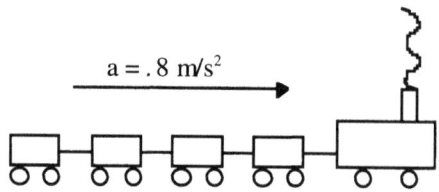

Newton's 2nd Law applies to a collection of objects as well as to individual objects. Thus, for the four cars (m_{total}) a = $F_{go} - F_{opp}$

$(2kg)(.8m/s^2) = 5N - f_k \Rightarrow f_k = 5N - 1.6N$ so $f_k = 3.4N$

☐☐ **A 50 kg box is sliding up an incline that makes an angle of 20° with respect to the horizontal. If $\mu_k = .25$, and the speed of the box at the bottom of the incline is 10 m/s, how far up the incline will the box slide before coming to rest?**

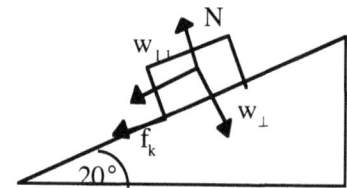

$v^2 = v_0^2 + 2ax$

Thus, if one can find the acceleration that acts on the box as it slides up the incline, the distance the box slides before coming to rest can be found.

$F_{un} = ma = F_{go} - F_{opp}$
$ma = -f_k - w_{\parallel}$

$ma = -\mu_k N - w_{\parallel}$

$a = \dfrac{-\mu_k mg \cos 20° - mg \sin 20°}{m}$

$a = -.25(9.8 m/s^2)\cos 20° - 9.8 m/s^2(\sin 20°)$

$a = -5.65 \text{ m/s}^2$

Thus, $0^2 = (10m/s)^2 + 2(-5.65m/s^2)x$
$\quad x = 8.85m$

☐☐ **A boy of mass 30kg and a girl of mass 20 kg are connected by a light rope as in the figure. If a horizontal force of 200N acts on the boy, (boy and girl are on a frictionless frozen lake), find (a) the acceleration and (b) the tension in the rope connecting the boy and girl.**

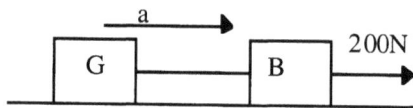

(a) Applying Newton's 2nd Law to the boy and girl,
$F_{un} = ma = F_{go} - F_{opp}$

$(50 \text{ kg})a = 200\text{N} \Rightarrow a = 4 \text{ m/s}^2$

(b) To find tension on the rope between the boy and girl, apply Newton's 2nd Law to the girl:
$F_{un} = ma = F_{go} - F_{opp}$

$(20\text{kg})\ 4\text{m/s}^2 = T - 0 \Rightarrow T = 80\text{N}$

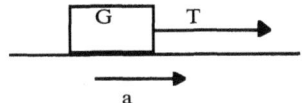

⬜⬜ A Saturn V rocket has a mass of 2.7×10^6 kg and a thrust (go force) of 3.3×10^7 N. Find the initial acceleration of the rocket.

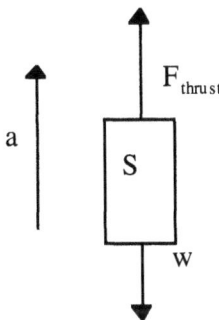

$f_{un} = ma = F_{go} - F_{opp} = F_{thrust} - w$

$(2.7 \times 10^6 \text{ kg})a = 3.3 \times 10^7 \text{ N} - (2.7 \times 10^6 \text{ kg})9.8\text{m/s}^2$

$a = 2.42 \text{ m/s}^2$

⬜⬜ A 9 kg block is supported with two ropes, as in the figure. Find the tension in each rope.

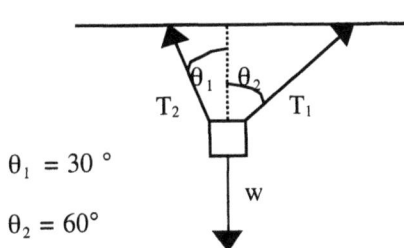

$\theta_1 = 30°$

$\theta_2 = 60°$

$w = mg = 88.2\text{N}$

$\Sigma F_{up} = \Sigma F_{down}$

$T_2\cos\theta_1 + T_1\cos\theta_2 = w \Rightarrow .866T_2 + .5T_1 = 88.2N$

$\Sigma F_{left} = \Sigma F_{right}$

$T_2\sin\theta_1 = T_1\sin\theta_2 \Rightarrow .5T_2 = .866T_1$

$\Rightarrow \dfrac{.5T_2}{.866} = T_1 = .577T_2$

Thus, $.866T_2 + .5(.577T_2) = 88.2N$

$$1.15T_2 = 88.2N \Rightarrow T_2 = 76.4N$$

$T_1 = \dfrac{.5T_2}{.866} = \dfrac{.5(76.4N)}{.866} \Rightarrow T_1 = 44.1N$

❑❑ Frances the talking mule refuses to pull a heavily loaded wagon because if she pulls hard on the wagon it pulls back on her with a force equal in magnitude but opposite in direction. Frances' story is that the harder she pulls on the wagon the harder it pulls back on her, so she won't even try. How does one counter her argument?

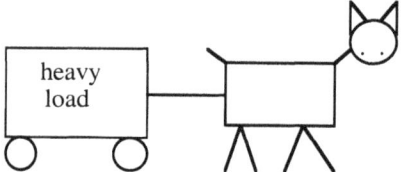

Action-reaction pairs (Newton's 3rd Law) never act on the same object.

❑❑ An object falling through the air reaches a constant velocity called the terminal velocity. How can this be?

If velocity is constant, then that means the weight is counter-balanced by the friction of the air opposing the motion of the object. The velocity for which the frictional force provided by the atmosphere equals the weight is called the terminal velocity.

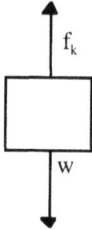

❑❑ Atwood's machine consists of two masses ($m_2 > m_1$) connected by a cord placed over a frictionless pulley as in the figure. (a) How could this device be used to find the acceleration due to gravity? (b) Find the tension in the cord supporting m_2 while m_2 is descending.

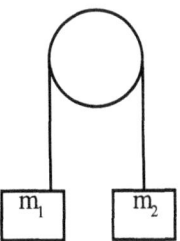

(a) for the system, Newton's 2nd Law says that
$F_{un} = (m_{system})a_{system} = F_{go} - F_{opp}$

$(m_1 + m_2)a = m_2g - m_1g \Rightarrow g = \left(\dfrac{m_1 + m_2}{m_2 - m_1}\right)a$

Hence, if one can find a one can find g.

(b) for m_2; $m_2g - T = m_2a \Rightarrow T = m_2g - m_2a = m_2(g-a)$

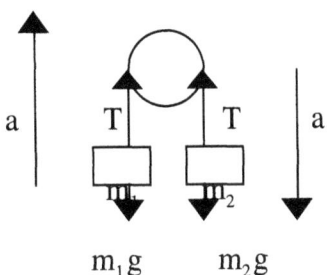

NEWTON'S LAWS II

THE FUNDAMENTAL IDEAS HERE ARE THE FOLLOWING:

- Radial acceleration comes about if an object is executing uniform circular motion (circular motion with constant speed).

- The magnitude of the radial acceleration is given by $a_{radial} = v^2/r$

- Since $F = ma$, $F_{radial} = (mv^2)/r$

- The direction of the radial force depends on the reference frame:

a) if the observer is in the at rest (inertial) reference frame the radial force is directed inward and is called centripetal force and has magnitude $(mv^2)/r$

b) if the observer is in the rotating (non-inertial) reference frame, then the radial force is directed outward and is called centrifugal force and has magnitude $(mv^2)/r$

- Newton's Law of Gravitation says that the force of attraction between any two objects is the universe is given by $F_{att} = \dfrac{Gm_1 m_2}{r^2}$, where G = gravitational constant, m_1 and m_2 are the masses of the two objects, and r is the distance between the two objects. In the case of extended bodies, the center of mass (that point which acts as if all of the mass were there) is at the geometric center.

- For satellite motion, the speed of the satellite in a circular orbit is given by
$v_s^2 = \dfrac{Gm_p}{r_s}$ where m_p is the mass of the planet about which the satellite is orbiting, G is the gravitational constant, and r_s is the radius of the satellite's orbit.

SOLVED PROBLEMS

☐☐ An object of mass 6 kg is moving in a circle of radius 5m, with a speed of 10 m/s. Find the magnitude of the radial acceleration.

$$a_{radial} = \dfrac{v^2}{r} = \dfrac{(10m/s)^2}{5m} = 20 m/s^2$$

☐☐ An object of mass 6kg is moving in a circle of radius 5m with a speed of 10 m/s. What is the magnitude and direction of the radial force seen by (a) the at rest observer, and (b) the observer in the moving reference frame?

(a) $F_{radial} = ma_{radial} = 120N$ in (toward the center of the circle)

(b) $F_{radial} = ma_{radial} = 120N$ out (away from the center of the circle)

❏❏ Given that the mass of the sun is 2 x 10³⁰kg, and the mass of the earth is 6 x 10²⁴kg, find the magnitude of the force of attraction between the earth and sun if the earth-to-sun distance is 1.49 x 10¹¹m.

$$F_{att} = \frac{Gm_e m_s}{r_{es}^2} = \frac{\left(6.67 \times 10^{-11} \frac{N-m^2}{kg^2}\right)\left(6 \times 10^{24} kg\right)\left(2 \times 10^{30} kg\right)}{\left(1.49 \times 10^{11} m\right)^2} = 3.61 \times 10^{22} N$$

❏❏ A 5kg object is moving in a vertical circle of radius .5m as in the figure. At what speed would the tension in the cord be 0 at the top of the circle?

Since the object is in the rotating reference frame, the tension in the cord would be 0 when
T + mg = F_out ⇒ mg = F_out

$$\Rightarrow \frac{mv^2}{r} = mg$$

$v = \sqrt{rg} = \sqrt{(.5)9.8} m/s$

$v = 2.21 m/s$

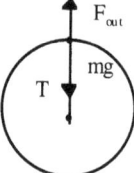

❏❏ For the previous problem, what would be the tension in the cord if the object were moving at 2 m/s at the bottom of the circle?

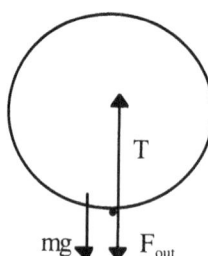

Since object is in rotating reference frame, then

$T = mg + F_{out} = mg + \dfrac{mv^2}{r}$

$T = (5kg)(9.8m/s^2) + \dfrac{(5kg)(2m/s)^2}{.5m}$

$T = 89N$

Note that for an at rest observer, the tension force T has to provide the inward (centripetal) force so that
$T = mg + F_{out}$ as before.

☐☐ **A car goes around a flat curve of radius 150m at a speed of 24m/s. What must the coefficient of friction be between the tires and the road for the car not to skid?**

If the car (which is in the rotating reference frame) just starts to slide sideways, then from the force conditions of equilibrium, $N = mg$ and $f_s = F_{out}$
Thus,

$u_s N = \dfrac{mv^2}{r} \Rightarrow u_s mg = \dfrac{mv^2}{r}$

and $u_s = \dfrac{v^2}{gr} = \dfrac{576 m^2/s^2}{(150m)9.8m/s^2} = .392$

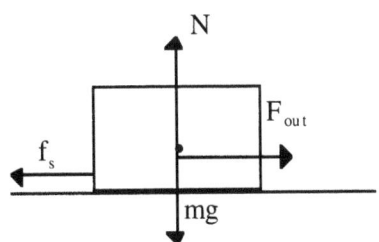

☐☐ **A pilot who weighs 700N is in an aircraft that is "Looping-the Loop", i.e., executing a circular maneuver in a vertical plane (radius = 500m) as in the figure. What force does the seat exert on the pilot if she is (a) moving at a speed of 100 knots (61.4m/s) at the top of the loop, and (b) moving at the bottom of the loop at 125 knots (64.3m/s)?**

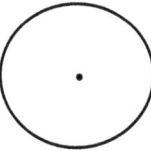

(a) The pilot's mass is 71.4kg (w = mg). At the top of loop, for the pilot who is in the rotating reference frame,

$$F_s + F_{out} = mg$$

$$F_s = mg - F_{out} = mg - mv^2/r$$

$$F_s = 700N - \frac{71.4kg(61.4m/s)^2}{500m}$$

$$F_s = 700N - 377N = 323N$$

(b) at the bottom of the loop,

$$F_s = F_{out} + mg$$

$$= \frac{mv^2}{r} + mg = \frac{(71.4kg)(64.3m/s)^2}{500} + 700N$$

$$F_s = 1290N$$

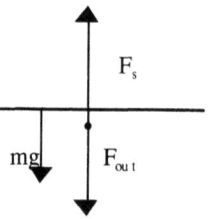

□□ A car goes around a curve of radius 90m. If a penny lies on a flat dash, at what speed would the penny start to slide if the coefficient of static friction is 0.5?

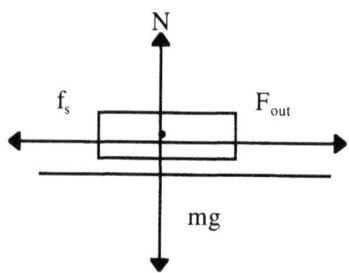

If the penny is just starting to slide, $\Sigma F_{up} = \Sigma F_{down}$
$$N = mg$$
and $\Sigma F_{left} = \Sigma F_{right}$
$$f_s = F_{out}$$

$$\Rightarrow u_s N = \frac{mv^2}{r}$$

$$\frac{u_s mgr}{m} = v^2 = .5(9.8m/s^2)90m$$

$$v^2 = 441 m^2/s^2$$

$$v = 21 m/s$$

❏❏ A centrifuge is rotating at 10,000 rpm. What is the magnitude of the radial acceleration that acts on a specimen that is 5 cm from the center of rotation? Express this acceleration as a multiple of g.

$a_{radial} = v^2/r$ \qquad\qquad 1 rev = $2\pi r$ = (6.28)(.05m) = .314m

$v = 10,000 \dfrac{rev}{min} \left(\dfrac{1 min}{60 s}\right)\left(\dfrac{.314m}{1 rev}\right) = 52.3 m/s$

$a_{radial} = \dfrac{(52.3 m/s)^2}{.05m} = 54,700 m/s^2 \left(\dfrac{1g}{9.8 m/s^2}\right) = 5580g$

❏❏ The moon orbits the earth in an almost circular orbit of radius 3.84×10^8 m and takes 27.4 days to complete one orbit around the earth. From this data find the mass of the earth.

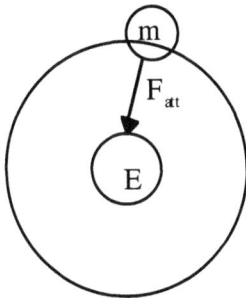

For the observer on the earth, the radial force is acting toward the earth.

$F_{att} = \dfrac{m_m v_m^2}{r_m} = \dfrac{G m_m m_e}{r_m^2} \Rightarrow v_m^2 = \dfrac{G m_e}{r_m}$

but for one trip around the earth $2\pi r_m = (v_m) t$

$v_m^2 = \dfrac{4\pi^2 r_m^2}{t^2}$

so that,

$\dfrac{4\pi^2}{t^2}\left(r_m^3\right) = G m_e$

$\dfrac{(4\pi^2)(3.84 \times 10^8 m)^3}{[(27.4) 86400 s]^2 \left(6.67 \times 10^{-11} \dfrac{N-m^2}{kg^2}\right)} = m_e$

$5.98 \times 10^{24} kg = m_e$

☐☐ Given that near the surface of the earth the weight of an object is mg, and that the radius of the earth is 6400km, find the mass of the earth from this data, assuming that the mass of the earth acts as if it were at its geometric center.

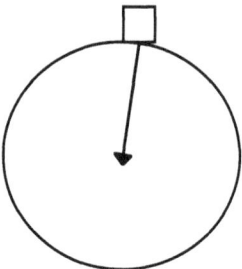

$$w = mg = F_{att} = \frac{Gm(m_e)}{R_e^2}$$

$$\Rightarrow \frac{(R_e^2)g}{G} = m_e = \frac{(6.4x10^6 m)^2 9.8 m/s^2}{6.67x10^{-11} \frac{N-m^2}{kg^2}}$$

$$m_e = 6.02x10^{24} kg$$

☐☐ At what angle should a curve (radius 120m) be banked so that if there is no friction a car would be able to stay on the road at 22 m/s if there were no friction?

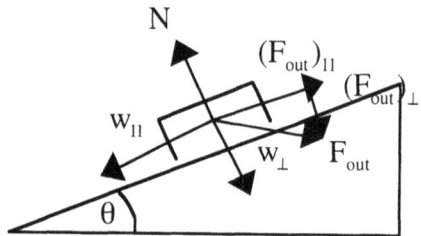

Here if the car is on the verge of sliding, then
$w_{11} = (F_{out})_{11}$

$$w\sin\theta = (F_{out})\cos\theta \Rightarrow mg\sin\theta = \frac{mv^2}{r}\cos\theta$$

$$\Rightarrow \tan\theta = \frac{v^2}{rg} = \frac{(22m/s)^2}{(120m)9.8m/s^2}$$

$$\theta = 22°$$

Thus, the road should be banked at an angle of 22°

⬜⬜ A block rests on an inclined plane as in the figure. Show that as the angle θ is increased, $\tan\theta_s = \mu_s$ where θ_s is that angle for which the block just slips.

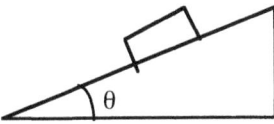

If the block is on the verge of slipping, $\Sigma F_{up} = \Sigma F_{down} \Rightarrow N = w_\perp$

$$\Sigma F_{left} = \Sigma F_{right} \Rightarrow w_{11} = f_s$$

Thus, $mg\sin\theta_s = \mu_s N$, but $N = mg\cos\theta_s$
so that $mg\sin\theta_s = \mu_s mg\cos\theta_s$

$\Rightarrow \dfrac{\sin\theta_s}{\cos\theta_s} = \mu_s = \tan\theta_s$

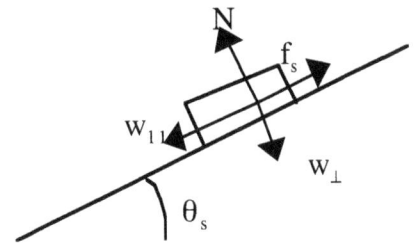

⬜⬜ The earth orbits the sun every 365.25 days in an almost circular orbit of radius 1.49×10^{11} m. From this data find the mass of the sun.

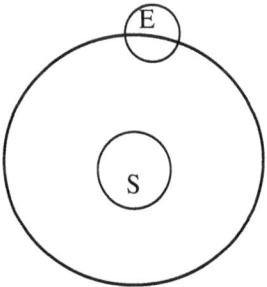

Since the earth orbits the sun, $F_{att} = \dfrac{m_e v_e^2}{r_e} = \dfrac{G m_e m_s}{r_e^2}$

Thus, $v_e^2 = \dfrac{G m_s}{r_e^2}$ but $v_e = \dfrac{2\pi r_e}{t}$

$\Rightarrow \dfrac{4\pi^2 r_e^3}{G t^2} = m_s = \dfrac{4\pi^2 (1.49 \times 10^{11} m)^3}{\left(6.6 \times 10^{-11} \dfrac{N-m^2}{kg^2}\right)(365.25 \, days)^2 (86,400 s/day)^2}$

$\Rightarrow m_s = 1.97 \times 10^{30}$ kg

☐☐ **Given that Mars has a mass of 6.34×10^{23} kg and a radius of 3.43×10^6 m, find the acceleration due to gravity on Mars.**

$$w_M = m_0 g_M = F_{att} = \frac{G m_0 m_M}{R_M^2}$$

$$g_M = \frac{G m_M}{R_M^2} = \frac{\left(6.67 \times 10^{-11} \frac{N-m^2}{kg^2}\right)\left(6.34 \times 10^{23} kg\right)}{\left(3.43 \times 10^6 m\right)^2}$$

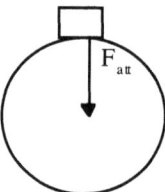

$g_M = 3.6 \text{ m/s}^2$

[Note that the assumption here is that all of the mass of Mars is at its center of mass.}

☐☐ **A person weights 800N on Earth. How much does this person weigh on Mars? (Use data from the previous question)**

$w_M = m_p g_M$ but $w_E = m_p g_E$
$\qquad\qquad\qquad\qquad 800N = m_p(9.8\text{m/s}^2)$
$\qquad\qquad\qquad\qquad \Rightarrow m_p = 81.6 \text{ kg}$

$\qquad = (81.6\text{kg})3.6\text{m/s}^2$

$w_m = 294N$

☐☐ **A proton (mass = 1.67×10^{-27} kg) moves in a circular path of 58 cm, with a speed of 2.5×10^7 m/s. Find (a) the time for one revolution, (b) the radial acceleration, and (c) the radial force as it appears to an at rest observer.**

(a) Disp = vt $\Rightarrow 2\pi r = vt$

$$\Rightarrow t = \frac{2\pi r}{v} = \frac{(6.28)(.58m)}{2.5 \times 10^7 m/s} = 1.46 \times 10^{-7} s$$

(b) $a_{radial} = \frac{v^2}{r} = \frac{(2.5 \times 10^7 m/s)^2}{.58m} = 1.08 \times 10^{15} m/s^2$

(c) $F_{radial} = m a^{radial} = (1.67 \times 10^{-27} \text{kg})(1.08 \times 10^{15} \text{m/s}^2)$

$F_{radial} = 1.8 \times 10^{-12}$ N directed in toward center of circular motion.

☐☐ An earth satellite has a circular orbit with a speed of 7.8 km/s. How long in minutes does it take the satellite to make one orbit around the earth? (mass of earth is 6x10²⁴kg)

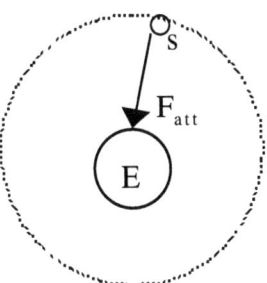

$$F_{att} = \frac{m_s v_s^2}{r_s} = \frac{G m_s m_e}{r_s^2}$$

$$v_s^2 = \frac{G m_e}{r_s} \Rightarrow r_s = \frac{G m_e}{v_s^2}$$

$$= \frac{\left(6.67 \times 10^{-11} \frac{N-m^2}{kg^2}\right)\left(6 \times 10^{24} kg\right)}{\left(7800 m/s^2\right)^2} = 6.58 \times 10^6 m$$

Thus, Disp = vt

2πr = vt

$$t = \frac{2\pi r}{v} = \frac{(6.28)(6.58 \times 10^6 m)}{7800 m/s} = 5296 s$$

⇒ time for orbit = 88.3 min

☐☐ A 5 kg block is attached to a vertical rod by two strings as in the figure. (a) If the system is rotated, how many revolutions per minute must the system make so that the tension in the upper cord is 80 N? (b) What is then the tension in the lower cord?

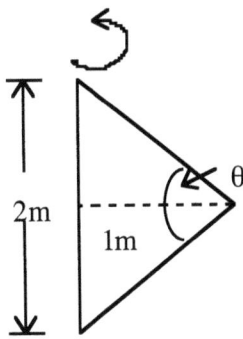

Apply the Force conditions of Equilibrium (looking at object that is rotating)

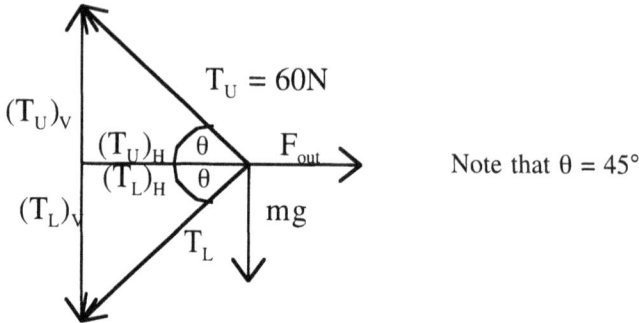

Note that θ = 45°

$\Sigma F_{up} = \Sigma F_{down}$

(1) $(T_U)_V = mg + (T_L)_V$

$\Sigma F_{left} = \Sigma F_{right}$

(2) $(T_U)_H + (T_L)_H = F_{out}$

(1') $(T_U)\sin\theta = mg + T_L\sin\theta$

$(80N)\sin 45° - mg = T_L\sin\theta$

$7.57N = T_L\sin 45°$

$T_L = 10.7N$

(2') $T_U\cos 45° + T_L\cos 45° = (mv^2/r)$

$(80N)(\cos 45°) + (10.7N)(\cos 45°) \dfrac{(5kg)v^2}{1m} \Rightarrow v = 3.58$ m/s

$v = 3.58 \dfrac{m}{s}\left(\dfrac{1 rev}{6.28m}\right)\left(\dfrac{60s}{1 min}\right) \Rightarrow v = 34.2 \dfrac{rev}{min}$

[note: to convert to rev/min, 1 rev = $2\pi r$ = 6.28 m]

WORK, POWER, AND ENERGY

THE FUNDAMENTAL IDEAS HERE ARE THE FOLLOWING:

- Work = Frcosθ, where F is magnitude of the force, r is magnitude of displacement, and θ is the angle between the force and the displacement. The work is zero if the angle is 90°, the displacement is zero, and/or the magnitude of the force is zero.

- $(Power)_{av} = \dfrac{work\ done}{time} = Fv_{av}$, since displacement divided by time is average velocity (the assumption here is that θ = 0°

- Energy is defined as the ability or capacity to do work.

- Kinetic energy is the energy that an object possesses because of its motion. Kinetic energy is denoted by the symbol K, and is given by $1/2\ mv^2$.

- Potential energy is the energy that an objecat possesses because of its position, and is denoted by the symbol U.

- Gravitational potential energy is given by U_g = mgh, where h is the distance above (or below) an arbitrarily chosen zero of gravitational potential energy.

- Elastic potential energy is the energy stored in a spring or other elastic object and is given by $U_s = 1/2\ k(\Delta x)^2$, where k is the spring constant (or force constant) of the spring, and Δx is the amount that the spring has been stretched or compressed. (Note that Hooke's Law says that F = kx, where F is the magnitude of the Force used to stretch a spring, k is the spring constant or force constant, and x is the amount the spring has been stretched or compressed.

- The total mechanical energy is the sum of the kinetic, gravitational potential, and elastic potential energies, i.e. $E = k + U_g + U_s$.

- The work energy theorem says that ΔK = WORK DONE.

- A more general form of the work energy theorem says that ΔE = WORK DONE.

- Friction is said to be a non-conservative force because it causes mechanical energy to change to heat and/or sound energy.

- If an object does work, it loses energy; if work is done on an object, it gains energy.

- The principal of the conservation of energy says that energy can change form, but it can neither be created nor destroyed.

- The definition of power is $(Power)_{av} = \dfrac{work\ done}{time}$, by using the definition of ΔE as work done, power can be rewritten as $(Power)_{av} = \dfrac{\Delta E}{time}$.

SOLVED PROBLEMS

❑❑ The force in the figure is applied to the block, which moves 8m. Find the work done by (a) force F and (b) the weight of the block.

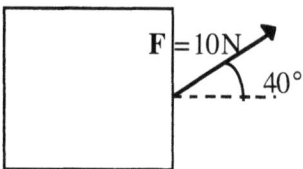

(a) Work = F(rcosθ)
   = (10N)(8m)cos 40°
   Work done by force F = 61.3N

(b) Here, work = F(rcosθ)
   = (w)(9.8m/s²)cos90°
   = 0
   So the work done by the weight force is 0.

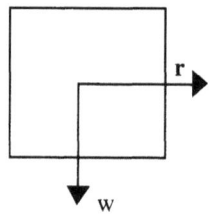

❑❑ An object is moving at 8 m/s and has a mass of 5kg. What is its kinetic energy?

k = 1/2(mv²) = 1/2(5kg)(8m/s)² = 160(kg-m²)/s²
k = 160J

❑❑ For the previous problem, how much work would the object do in coming to rest?

From Δk = work done, the object would do 160J of work.

❑❑ A spring has a spring constant of 35N/m. How much force is necessary to stretch the spring 12 cm? What energy is then stored in the spring?

F = kx ⇒ (35N/m)(.12m) = F = 4.2N
U_s = 1/2(k)(Δx)² = 1/2(35N/m)(.12m)² = .252J

❑❑ A 14kg object rests on a table that is 4m high. Find the gravitation potential energy when the 0 of U_g is at (a) the floor, and (b) the tabletop.

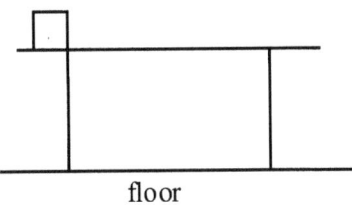
floor

(a) $U_g = mgh = (14kg)(9.8m/s^2)(4m) \approx 549J$

(b) $U_g = mgh = 0$

❑❑ A small engine develops 4 H.P. What is the maximum load that this engine could lift at 3 m/s as in the figure. (Ignore friction) (1 H.P. = 746 J/s)

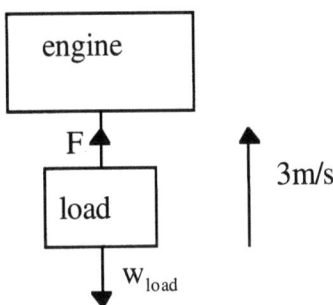

$P = Fv \Rightarrow \dfrac{746 J/s}{1 H.P.}(4 H.P.) = F(3m/s)$

$\Rightarrow F = 995N = w_{load}$ if load is going up at constant velocity.

❑❑ An object (mass = 4kg) starts from rest and slides down the incline as in the figure. What is the speed of the object at the bottom of the incline? (Ignore friction)

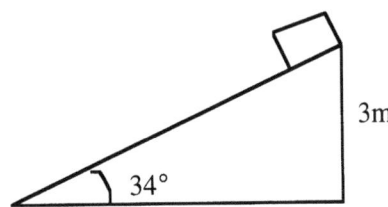

Place 0 of U_g at bottom of incline. If there is no friction, then
$E_{top} = E_{bottom}$

$(U_g)_{top} = (k)_{bottom}$

mgh = (1/2) mv²
2gh = v = $\sqrt{2(9.8 m/s^2)(3m)}$
v = 7.67 m/s

□□ For the previous problem, let there be friction. Find the work done by the frictional force as the block slides down the incline and the magnitude of the frictional force, if the speed of the block at the bottom is 3 m/s.

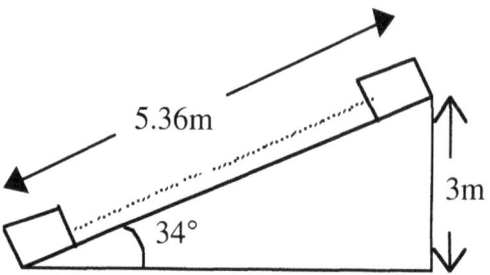

ΔE = work done
If $U_g = 0$ is at the bottom of the incline, then ΔE = work done (Note that block slides 5.36m down the incline)

$E_{bottom} - E_{top}$ = work done
(1/2)mv² - mgh = work done

(1/2)[4kg](9m²/s²) - 4kg(9.8m/s²)(3m) = work done
                                    - 99.6J = work done

Since, ΔE = work done
    -99.6J = $(f_k)(r)\cos 180°$

$\Rightarrow \dfrac{+99.6J}{5.36m} = f_k = 18.6N$

□□ The XF-4 experimental aircraft has design specifications such that the aircraft starts from rest and 5 minutes later is flying in level flight at an altitude(height) of 10,000 m above the runway at 600 knots (309m/s). Ignoring friction, what size engine (in H.P.) would be necessary if the mass of the aircraft is 3500 kg?

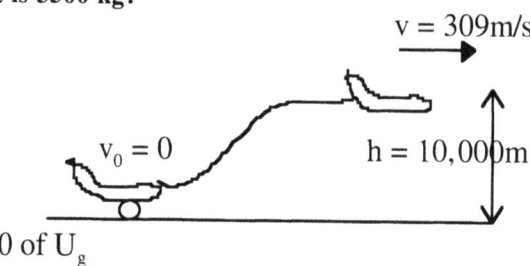

$P = \dfrac{\Delta E}{time}$

$P_{av} = \dfrac{E - E_0}{time} = \dfrac{mgh + \dfrac{1}{2}mv^2}{time}$

$$= \frac{(3500kg)(9.8m/s^2)(10,000m) + \frac{1}{2}(3500kg)(309m/s)^2}{300s}$$

$$P_{av} \cong 1.7 \times 10^6 J/s \left(\frac{1 H.P.}{746 J/s}\right) \cong 2280 H.P.$$

❏❏ In the previous problem, if only 30% of the output of the engine were available to do work because of friction, what size engine (in H.P.) would be required?

$$P_{out} = .3 P_{in}$$
$$\Rightarrow P_{in} = \frac{2280 H.P.}{.3} \cong 7600 H.P.$$

❏❏ If it takes 40 H.P. to drive a bus at a constant 22 m/s, what is the magnitude of the retarding (friction) force that acts on the bus?

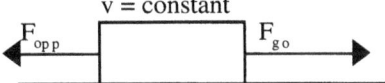

If the velocity is constant, $F_{go} = F_{opp} \Rightarrow P = F v_{av}$ so that

$$(40 H.P.)\left(\frac{746 J/s}{1 H.P.}\right) = (F_{go})(22 m/s)$$

$$\Rightarrow F_{go} = 1356 N$$

Thus, the total retarding force is 1356N

❏❏ How long would it take a 4 H.P. engine to lift at constant speed an 800N person a distance of 10m? (See figure)

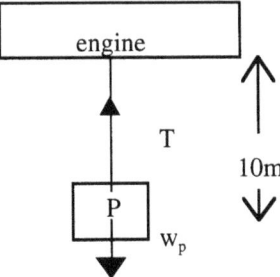

$$P = \frac{\Delta E}{time} \Rightarrow time = \frac{\Delta E}{T}$$

$$time = \frac{mgh}{P}$$

$$time = \frac{(800N)(10m)}{(4 H.P.)(746 J/s)} = 2.68s$$

☐☐ **A 2kg mass is dropped from a height of 2m above the end of a vertical spring with a force constant of 60N/m. How far will the spring be compressed when the mass is brought to rest? [Hint: This requires the solution to a quadratic equation]**

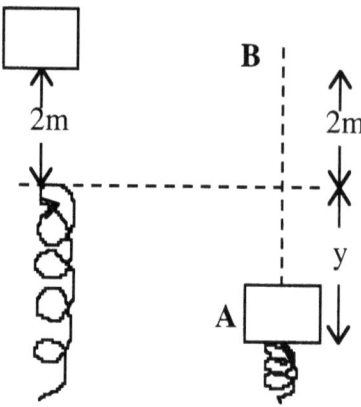

Place the 0 of U_g at point **A**. Then $E_A = E_B$
$(1/2)ky^2 = mg(y + 2m)$
$(1/2)(60N/m)y^2 = 19.6N(y + 2m)$
or $30y^2 - 19.6y - 2 = 0$

so that $y = .743m$ or $y = -.09m$

the spring would compress .743m, since 0 of U_g is at A.

☐☐ **A 100g ball is thrown up with an initial speed of 27m/s and rises to a height of 30m. What work was done by resistive forces (friction, etc)?**

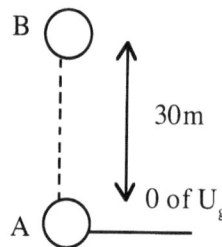

ΔE = work done
$E_B - E_A$ = work done
$mgh - (1/2)mv_A^2$ = work done
$29.4J - 45J$ = work done
$-15.6J$ of work done by resistive forces.

☐☐ **In the previous problem, if the resistive forces do -16 Joules of work on the ball as it travels down from B to A, what would be the speed of the ball when it returns to point A?**

ΔE = work done   (0 of U_g at point **A**)
$E_A - E_B = -16J$
$(1/2)mv_A^2 = -16J + mgh$
$(1/2)mv_A^2 = -16J + 29.4J$
$v_A^2 = \dfrac{(13.4J)(2)}{.1kg} \Rightarrow v_A = 16.4 m/s$

☐☐ After an accident a car leaves skid marks 40m long in sliding to rest. The driver says that she was not speeding (the speed limit is 40 mph, or 17.9 m/s). Experiment shows that the coefficient of kinetic friction between the tires and the pavement is 0.7. Was the driver speeding?

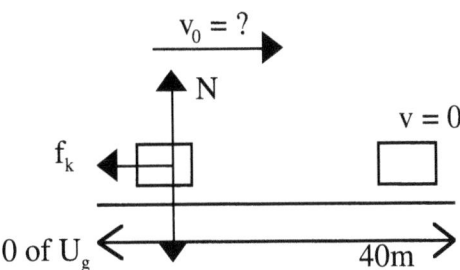

ΔE = work done
Placing the 0 of U_g at the pavement
$E - E_0$ = work done

$0 - (1/2)mv_0^2 = f_k r \cos\theta$
$\quad - (1/2)mv_0^2 = (\mu_k mg) r (-1)$

$\quad v_0^2 = 2\mu_k g r$
$\quad\quad = 2(.7)(9.8 m/s^2) 40m$
$\quad v_0^2 = 549 m^2/s^2$

$\quad v_0 = 23.4$ m/s

Yes, the driver was speeding.

☐☐ A 15 kg block starts from rest and is pushed 15m up an inclined plane that is inclined at an angle of 40° with the horizontal by a force F of 200N acting parallel to the plane. Given that the coefficient of kinetic friction between the block and the plane is 0.2, find (a) the work done by force F, (b) the increase in the potential energy of the block, (c) the work done by the friction force, (d) the kinetic energy of the block, (e) the speed of the block after force F has pushed the block 15m.

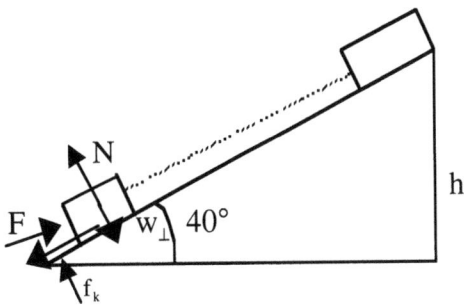

(a) work done by force F = $Fr\cos\theta$ = 3000 joules

(b) $\Delta U_g = U_g - (U_g)_0 = mgh$ [if 0 of U_g at bottom of plane]
$\quad\quad = (15 kg)(9.8 m/s^2) 9.64 m$
$\quad\quad = 1420 J$

(c) work done by the friction force is found by work = $(f_k)r\cos\theta$
$f_k = \mu_k N = \mu_k mg\cos\theta = (.2)(15kg)(9.8m/s^2)\cos 40°$
$= 22.5N$
Thus, work done by the friction force = $(22.5N)(15m)\cos 180° = -338J$

(d) from ΔE = work done
$(k + U_g) - (k_0 + (U_g)_0)$ = work done
$k + 1420J = 3000J - 338J$
$k = 1240J$

(e) $k = (1/2)mv^2 \Rightarrow 1240J = (.5)(15kg)v^2$
$166 m^2/s^2 = v^2$
12.9 m/s $= v$

□□ The spring of a spring gun has a force constant of 400 N/m. It is compressed 0.06m and a pellet of mass 0.01 kg is placed in the barrel against the compressed spring. If the spring gun is horizontal, find the speed with which the pellet leaves the gun if (a) there is no friction and (b) a constant retarding force of 10N acts on the pellet as it travels down the 6 cm length of the barrel.

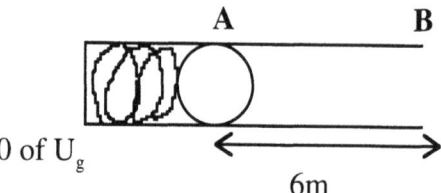

(a) If there is no friction, $E_A = E_B$
$(1/2) k(\Delta x)^2 = (1/2)mv^2$

$$\frac{\left(400 \frac{N}{m}\right)(.06m)^2}{.01kg} = v^2 = 144 m^2/s^2$$

$\Rightarrow v = 12$ m/s

(b)

here, ΔE = work done

$E_B - E_A = (F_R)r\cos 180°$

$(1/2) mv^2 - (1/2)k(\Delta x)^2 = (F_R)(r)(-1)$

$(1/2)mv^2 = (1/2)k(\Delta x)^2 - (F_R)r$

$v^2 = \dfrac{(200N/m)(.06m)^2 - 10N(.06m)}{.5(.01kg)} \Rightarrow v = 4.9 m/s$

❏❏ A 15kg object is moving at 14m/s when it is 4 m above the ground. Find its total mechanical energy (a) relative to a 0 of U_g at the ground, and (b) relative to a 0 of U_g at the same level as the object.

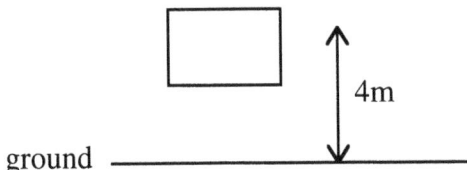

(a) $E = k + U_g$

$E = (1/2)mv^2 + mgh$

$= (1/2)(15kg)(14m/s)^2 + 15kg(9.8m/s^2)(4m)$

$E_{total} = 1470J + 588J = 2058J$ for 0 of U_g at the ground.

(b) For 0 of U_g at the same level as the object, $E = k = 1470J$

❏❏ For the previous problem, how much energy must be dissipated (changed in form) if the object falls to the ground ($U_g = 0$ at ground level) and collides with the ground?

ΔE = work done $\Rightarrow 0 - E_0$ = work done
-2058J = work done against dissipative forces.

IMPULSE, MOMENTUM AND COLLISIONS

THE FUNDAMENTAL IDEAS HERE ARE THE FOLLOWING:

- Rewriting Newton's Second Law, F = ma, gives $F = \dfrac{m(v - v_0)}{t}$
  or $F_{av} t = mv - mv_0$, where F_{av} is an average force over the time t.

- $F_{av} t$ is called impulse and has units of N-s (or kg-m/s)
  mv is called linear momentum, P, and has units kg-m/s (or N-s)

- Thus, $\vec{F}_{av} t = \Delta \vec{P}$, or, impulse equals change in momentum. Note again that t is the time the average force acts.

- If F is 0, that it if the external force is 0 when two or more objects collide, then
  $0 = \Delta \vec{P}$ which implies that $\vec{P} = \vec{P}_0$, that is in a collision, momentum is conserved.

- A rocket (as well as a jet engine) is a reaction engine, because the thrust force (go force) is the reaction to the force that acts on gases and combustion products that are ejected from the rocket.

- The thrust force for a rocket is given by $F_{thrust} = -v_{ex} \dfrac{\Delta m}{\Delta t}$, where v_{ex} is exhaust gas velocity, relative to the rocket, and $\dfrac{\Delta m}{\Delta t}$ is change in mass with respect to time.
  [The - is because the rocket is losing mass, i.e., $\dfrac{\Delta m}{\Delta t}$ is -]

- There are three types of collisions:
  - inelastic
  - perfectly elastic
  - elastic

- The inelastic collision is characterized by (a) the objects stick together after the collision and move off with a common velocity, and (b) the momentum before the collision is the momentum after the collision. $(\vec{P}_o = \vec{P})$.

- The perfectly elastic collision is characterized by (a) the momentum before the collision is the momentum after the collision $(\vec{P}_0 = \vec{P})$ (b) the kinetic energy of the system before the collision is the kinetic energy of the system after the collision ($k_0 = k$), and (c) the relative velocity before the collision is the negative of the relative velocity after the collision, i.e. $v_1 - v_2 = -(v_1' - v_2')$ where ' denotes after collision.

- The elastic collision lies between the inelastic collision and the perfectly elastic collision.

  Here, (a) momentum before the collision is the momentum after the collision, $\vec{P}_0 = \vec{P}$, and (b) the relative velocity of the objects after the collision is a fraction of the relative velocity before the collision, i.e., $e(v_1 - v_2) = -(v_1' - v_2')$. Here e is called the coefficient of restitution and is a number between 0 and 1, i.e. $0 < e < 1$.

- Force, impulse and momentum are vectors. If two vectors are equal, then their x components are equal, and their y components are equal.

SOLVED PROBLEMS

❑❑ **A 2kg object moving north at 2 m/s has what linear momentum?**

p = mv ⟹ $\vec{p}$ = (2kg)(2m/s) = 4 kg - m/s , north.

❑❑ **For the previous question, what force would be necessary to bring the object to rest in 0.4 s?**

Impulse = change in momentum

$(\vec{F}_{av})t = m\vec{v} - m\vec{v}_0 \Rightarrow \vec{F}_{av} = \dfrac{0 - 4kg - m/s}{.4s} = -1N$

or $\vec{F}$ to bring object to rest is 1 N South

❑❑ **An object of mass m is moving with speed v. (a) By what factor is the momentum increased if the speed increases by a vector of 2? (b) By how much is the kinetic energy increased?**

(a) p_0 = mv
    p = m(2v) = $2p_0$ ⟹ momentum is increased by a factor of 2.
(b) $k_0 = (1/2)mv^2$
    $k = (1/2)(2v)^2 \Rightarrow k = 4k_0 \Rightarrow$ kinetic energy is increased by a factor of 4.

❑❑ **A car (mass 1350 kg) going east at 25 m/s has a head-on inelastic collision with a truck (mass 1900 kg) going west at 20 m/s. Find the final velocity of the vehicles after the collision.**

Momentum is conserved. Thus, $m_c v_c + m_T v_T = (m_c + m_T)v'$
(1350kg)25m/s + 1900kg(-20m/s) = (1350kg + 1900kg)v'
-1.31 m/s = v'

Thus, the common velocity after the collision is 1.31 m/s west.

❑❑ **A bullet (mass 25g) travelling at 850 m/s collides with the trunk of a tree and penetrates to a depth of 45 cm. Find (a) the change in momentum of the bullet, (b) the impulse that acts on the bullet, (c) the average force that acts on the bullet to bring it to rest.**

(a) mv - mv_0 = 0 - .025 kg(850 m/s)
              = -21.3 kg-m/s.

(b) impulse = change in momentum, so
    impulse = -21.3 N-s

(c) $\begin{bmatrix} x = v_{av}t \\ \dfrac{.45}{425 m/s} = t = 1.06 \times 10^{-3} s \end{bmatrix}$;

Thus, $F_{av} t$ = impulse ⟹ $F_{av} = \dfrac{-21.3 N-s}{1.06 \times 10^{-3} s} = -2.01 \times 10^4 N$

☐☐ For a ball dropped from a height h_1 above a floor that rebounds to a height h_2, show that the coefficient of restitution squared is the ratio of the heights ($e^2 = h_2/h_1$).

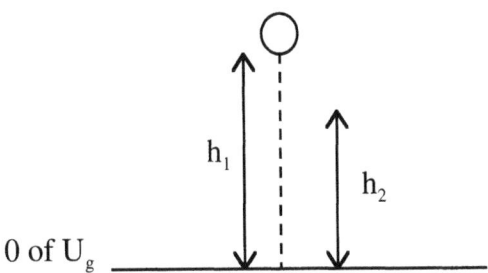

Since $e = \dfrac{-v_1'}{v_1}$; $e^2 = \dfrac{(v_1')^2}{v_1^2}$

using an energy approach: before the collision
$E_{top} = E_{bottom}$
$mgh_1 = (1/2)mv_1^2 \Rightarrow 2gh_1 = v_1^2$

after the collision
$E_{bottom} = E_{top}$
$(1/2)m(v')^2 = mgh_2 \Rightarrow (v')^2 = 2gh_2$

Thus, $e^2 = \dfrac{2gh_2}{2gh_1} = \dfrac{h_2}{h_1}$

☐☐ A rocket develops 10,000N of thrust and has an exhaust gas velocity of 300m/s. What is the mass ejected each second? (What is $(\Delta m)/(\Delta t)$?)

$F_{thrust} = -v_{ex}\dfrac{\Delta m}{\Delta t}$

$\Rightarrow 10,000N = -300\dfrac{m}{s}\left(\dfrac{\Delta m}{\Delta t}\right)$

$\dfrac{\Delta m}{\Delta t} = -33.3\dfrac{N}{m/s} = -33.3\dfrac{kg}{s}$

☐☐ A ball is dropped from a height of 3m and rebounds from the floor to a height of 2m. Find the coefficient of restitution for the ball-floor collision.

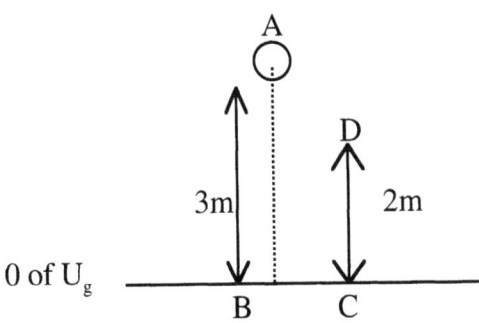

To find v_1 (velocity of ball before the collision):
$U_A = k_B \Rightarrow mgh_A = 1/2 m$

$v_1 = \sqrt{2gh_1} = -7.67 m/s$

To find v_1' (velocity of ball after the collision; $U_D = k_C$, thus

$v_1' = \sqrt{2gh_2} = 6.26 m/s$

For an elastic collision, $v_1 - v_2 = -e(v_1' - v_2') \Rightarrow e = \dfrac{-v_1'}{v_1}$

Thus, $e = \dfrac{-6.26 m/s}{-7.67 m/s}$

$e = .816$

☐☐ **A 10g bullet is fired horizontally into a 4kg wood block at rest on a horizontal surface where $\mu_k = 0.2$. The bullet remains embedded in the block, which is observed to slide 30cm before coming to rest. Find the initial velocity of the bullets.**

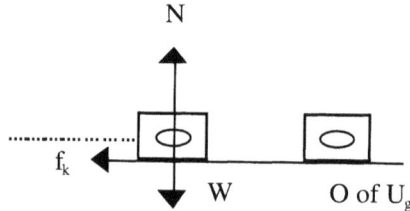

Here, momentum is conserved.
$(mv)_0 = mv$

$m_b v_b = (m_b + m_{bL})v' \Rightarrow v_b = \dfrac{(m_b + m_{bL})v'}{m_b}$

If block slides 30cm, from the work energy theorem, ΔE = work done
$0 - (1/2)(m_b + m_{bL})v'^2 = (f_k)(r)\cos 180°$
$-1/2(m_{b+m}bL)v'^2 = \mu_k(m_b + m_{bL})gr(-1)$
so that $(v')^2 = 2\mu_k gr = 2(.2)(9.8 m/s^2)(.3m)$
$(v')^2 = 1.18 m^2/s^2$
$v' = 1.08 m/s$

$v_b = \dfrac{(.01kg + 3kg)}{.01kg}(1.08 m/s)$

$v_b \approx 325 m/s$

☐☐ **A steel marble (mass = .4kg) falls from a height of 5m onto a steel plate and rebounds to its original height. Find (a) the impulse that acts on the ball during the impact, and (b) the average force that acts on the ball if the time of collision is 0.005s.**

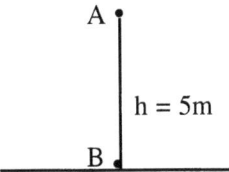

(a) If the ball rebounds to its initial height, $v = v_0$
To find the v_0 of the ball before the collision, $E_A = E_B \Rightarrow$
$v_B = \sqrt{2gh} = v_0$
$v_0 \cong -9.9 m/s$ and $v = 9.9 m/s$

Since impulse = change in momentum
impulse = $mv - mv_0$ = .4kg(9.9m/s) - (.4kg)(-9.9m/s)
impulse = 7.92 N-s.

(b) Thus, $(F_{av})t$ = impulse
so that $F_{av} = \dfrac{impulse}{time} = \dfrac{7.92 N-s}{.005s}$

$F_{av} \cong 1580N$

☐☐ **A 10g block moving with a velocity of 25cm/s to the right has a head-on (one-dimensional) collision with a 30g block moving to the left with a velocity of 15 cm/s. Find the velocity of each block if (a) the collision is inelastic and (b) the collision is perfectly elastic.**

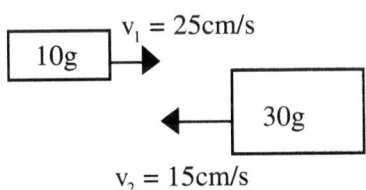

(a) If the collision is inelastic, then the blocks stick together after the collision. Since momentum is conserved,
$m_1v_1 + m_2v_2 = (m_1 + m_2)v'$
$\Rightarrow v' = \dfrac{m_1v_1 + m_2v_2}{m_1 + m_2}$

$= \dfrac{(10g)25cm/s + 30g(-15cm/s)}{40g}$

v' = -5cm/s or 5cm/s to the left.

(b)

If the collision is perfectly elastic, then $m_1v_1 + m_2v_2 = m_1v_1' + m_2v_2'$
and $v_1 - v_2 = -(v_1' - v_2')$

Thus, $10g(25cm/s) + 30g(-15cm/s) = (10g)v_1' + 30gv_2'$
and $25cm/s - (-15cm/s) = -v_1' + v_2'$
or
$-200cm/s = 10v_1' + 30v_2'$
$40cm/s = -v_1' + v_2'$.

Solving these two equations gives that $v_1' = -35cm/s$ and $v_2' = 5cm/s$

❑❑ **A model rocket burns .08kg of fuel each second, ejecting it as a gas with a velocity relative to the rocket of 6,000m/s. At launch, the rocket weights 1000N. (a) Find the thrust developed by the rocket and (b) will this rocket lift off the launch pad?**

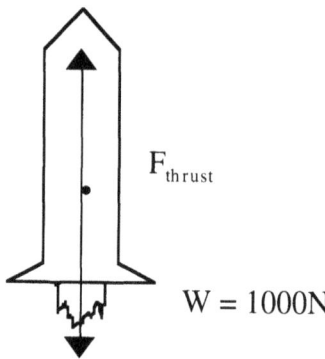

(a) $F_{thrust} = -v_{ex} \dfrac{\Delta m}{\Delta t} = -6000 m/s \left(-.08 \dfrac{kg}{s}\right) = 480N$

(b) weight = 1000N
thrust = 480N

Rocket won't lift off.

❑❑ **A person of mass 95kg, (standing on ice), fires a pistol horizontally The bullet has a mass of 15g and a velocity of 850m/s, what is the recoil velocity of the person? (Ignore friction)**

Here momentum is conserved.

$0 = m_bv_b + m_pv_p \Rightarrow v_p = \dfrac{-m_bv_b}{m_p} = \dfrac{-.015kg(850m/s)}{95kg}$

$v_p = -.134m/s$

Thus, the person acquires a velocity opposite of the bullet of magnitude .134m/s.

❑❑ A car of mass m moving a speed v has a head-on inelastic collision with an identical car moving at the same speed in the opposite direction. What force acts on the first car (and its occupants) because of the collision? (Assume time of collision is t.)

Momentum is conserved, thus $mv + m(-v) = (m + m)v' \Rightarrow v' = 0$.
Thus, $F_{av}t = mv - mv_0$

$$F_{av} = \frac{-mv_0}{t}$$

❑❑ For the previous problem, suppose that the collision were perfectly elastic. Find the force that acts on the first car.

Here momentum is conserved $\Rightarrow m_1v_1 + m_2(-v_2) = mv_1' + mv_2'$
$\Rightarrow v_1' = -v_2'$
$(1/2)mv^2 + (1/2)mv^2 = (1/2)m(v_1')^2 + (1/2)m(v_2')^2$
$\Rightarrow v^2 = (v_1')^2 \Rightarrow -v = v_1'$

so that $(F_{av})t = mv - mv_0 \Rightarrow F_{av} = \frac{2mv_0}{t}$

Thus, the force is greater for the perfectly elastic collision.

❑❑ Car A (m = 1350kg) is going east at 20m/s and has an inelastic collision with car B (mass 2020kg) going south at 15m/s. Ignoring friction, find the magnitude and direction of their common velocity after the collision.

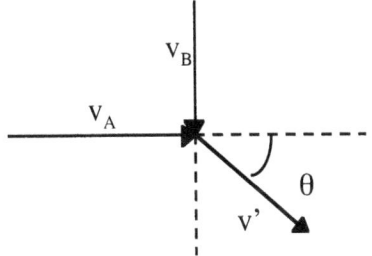

$\vec{p}_0 = \vec{p}$

East component
$m_A v_A = (m_A + m_B)v'\cos\theta$

South component
$m_B v_B = (m_A + m_B)v'\sin\theta$

and dividing the south component by the east component:

$$\Rightarrow \frac{(2020kg)(15m/s)}{(1350kg)(20m/s)} = \frac{(1350 + 2020kg)v'\sin\theta}{(1350 + 2020kg)v'\cos\theta}$$
$\Rightarrow 1.12 = \tan\theta \Rightarrow \theta = 48.2°$

so that, substituting in the equation for the south component gives
$(2020kg)(15m/s) = (3370kg)(v')\sin 48.2°$
$\Rightarrow v' = 12m/s$

After collision then, the velocity of the coupled cars is 12m/s 48.2° S of E.

☐☐ For the previous problem, what was the change in the kinetic energy of the cars because of the collision?

$\Delta k = k - k_0$

$\quad = \frac{1}{2}(m_A + m_B)v'^2 - \left[\frac{1}{2}m_A v_A^2 + \frac{1}{2}m_B v_B^2\right]$

$\quad = \frac{1}{2}(3370kg)(12m/s)^2 - \left[\frac{1}{2}(1350kg)(20m/s)^2 + \frac{1}{2}(2250kg)(15m/s)^2\right]$

$\Delta k = 2.43 \times 10^5 J - 5.23 \times 10^5 J$
$\Delta k = -2.8 \times 10^5 J$

☐☐ An α particle (mass 4u) moving at 1.8×10^7 m/s collides with a gold nucleus of mass 197u which is initially at rest. Find the velocities of the α particle and the gold nucleus after the collision. Assume that the collision is one dimensional and perfectly elastic.

$m_1 v_1 + m_2 v_2 = m_1 v_1' + m_2 v_2' \Rightarrow 4u(1.8 \times 10^7 m/s) = (4u)v_1' + (197u)v_2'$

$v_1 - v_2 = -(v_1' - v_2') \Rightarrow 1.8 \times 10^7 m/s = -v_1' + v_2'$

Thus, $7.2 \times 10^7 m/s = 4v_1' + 197 v_2'$

and $1.8 \times 10^7 m/s = -v_1' + v_2'$

solving the system of equations leads to:

$v_\alpha' = -1.73 \times 10^7 m/s$ and $v_{Au}' = 7.16 \times 10^5$ m/s.

☐☐ A machine gun fires 15g bullets with a velocity of 500m/s at a rate of 600 rounds/minute. Find (a) the impulse that acts on the bullet, (b) the average force that act on each bullet, and (c) the average force that acts on the person pulling the trigger.

(a) Impulse equals the change in momentum $\Rightarrow F_{av}\Delta t = mv - mv_0 = (.015kg)500m/s - 0$.
thus, impulse = 7.5N-s

(b) $F_{av} = \dfrac{impulse}{time} = \dfrac{600 rounds}{min} = \dfrac{60 rounds}{sec}$

The time that each bullet is in the barrel is (1/60)s.

Thus, $(F_{av})_{bullet} = \dfrac{7.5N-s}{\frac{1}{60}s} = 450N$

(c) Newton's Third Law tells us that $F_{gunman} = -F_{bullet} = -450N$

☐☐ A missile has a mass of 15,000kg on the launch pad. It is fired vertically by engines that develop 160,000 N of thrust, ejecting gasses at the rate of 80 kg/s. Find (a) the speed of the exhaust gasses relative to the rocket, (b) the acceleration of the rocket at the instant of launch, and (c) the acceleration of the rocket 50 s after launch.

(a) $F_{thrust} = -v_{ex}\dfrac{\Delta m}{\Delta t} \Rightarrow 160,000N = -v_{ex}\left(-\dfrac{80kg}{s}\right)$

$\Rightarrow v_{ex} = 2000 m/s$

(b) $F_{un} = F_{go} - F_{opp} = ma \Rightarrow (15,000kg)a = 160,000N - (15,000)9.8m/s^2$
a at launch = $.867 m/s^2$

(c) At t = 50s, the mass is no longer 15,000kg. Thus $\left(\frac{\Delta m}{\Delta t}\right)\Delta t = \Delta m$

$\left(\frac{80kg}{s}\right)(50s) = 4000 kg$

Thus, $ma = F_{go} - F_{opp}$
$(11,000kg)a = 160,000N - 11,000kg(9.8m/s^2)$
so $a = 4.75 m/s^2$ 50s after launch.

◻◻ If the coefficient of restitution for a golf ball is 0.8, find the height to which the ball rebounds if it is dropped from a height of 3m.

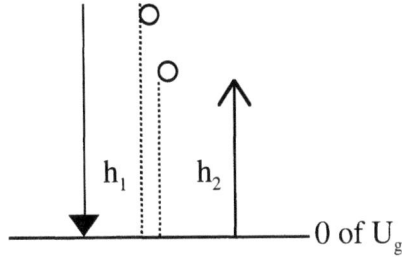

$e(v_1 - v_2) = -(v_1' - v_2') \Rightarrow e = -\frac{v_1'}{v_1} = \frac{\sqrt{2gh_2}}{\sqrt{2gh_1}} \Rightarrow e^2 = \frac{h_2}{h_1}$

so that $h_2 = e^2 h_1 = (.8)^2(3m) = 1.92m$

◻◻ A 15g bullet hits and stays in a 1.5kg block suspended at the end of a 1.4m string. (The block is initially at rest.) After the collision, the string moves to an angle of 30° to the vertical. Find the initial speed of the bullet.

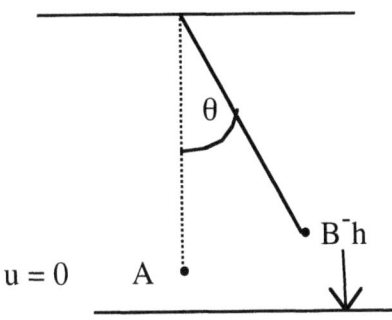

After collision, $E_A = E_B$

$$\frac{1}{2}(m_b + m_{bL})(v')^2 = (m_b + m_{bL})gh$$
$$v' = \sqrt{2gh}$$
$$ = \sqrt{2g(l - l\cos\theta)}$$
$$v' = \sqrt{19.6 m/s^2(1.4m - 1.4m\cos 30°)} = 1.92 m/s$$

Writing the conservation of momentum for the bullet-block collision,
$$m_b v_b = (m_b + m_{bL})v' \Rightarrow v_b = \frac{(1.515 kg)(1.92 m/s)}{.015 kg} = 194 m/s$$

ROTATIONAL MOTION I

THE FUNDAMENTAL IDEAS ARE THE FOLLOWING:

- Angular displacement θ can be measured in degrees, revolutions, radians or grads. The unit of angular displacement is chosen to be radians.

- Angular velocity w has a unit of rev/min or radians/second. The definition of average angular velocity is
$$w_{av} = \frac{\Delta \theta}{\Delta t}$$, with units of rad/s

- Angular acceleration α has units of radians/seconds. $\alpha_{av} = \frac{\Delta w}{\Delta t}$ is the definition of angular acceleration.

- If the angular acceleration is constant, the tools of motion apply. Thus,
  (1) $\theta = w_{av} t$
  (2) $w_{av} = \frac{w + w_0}{2}$
  (3) $w = w_0 + \alpha t$
  (4) $w^2 = w_0^2 + 2\alpha\theta$
  (5) $\theta = (1/2)\alpha t^2 + w_0 t$

- The definition of a radian is given by $\theta(rad) = \frac{S}{R}$

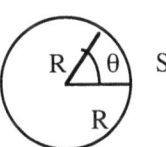

- Thus, $S = R\theta$ is the expression that relates linear displacement to angular displacement.
- $V = RW$ is the expression that relates linear velocity to angular velocity.
- $a_T = R\alpha$ is the expression that relates linear acceleration to angular acceleration (it is called tangential linear acceleration).
- In linear motion, $a_{radial} = \frac{v^2}{r}$
- In rotational motion, $a_{radial} = rw^2$
- For an object describing circular motion $\vec{a}_{total} = \vec{a}_T + \vec{a}_R$

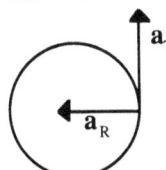

SOLVED PROBLEMS

❏❏ **Given that θ(rad) = S/R, how many radians are in one revolution?**

$$\theta = \frac{S}{R} = \frac{2\pi R}{R} = 2\pi$$

Thus, 1 rev = 2π rad

❏❏ **w = 1rev/min equals how many rad/s?**

$$\frac{1 rev}{min}\left(\frac{1 min}{60 s}\right)\left(\frac{2\pi rad}{1 rev}\right) = \frac{.1047 rad}{s}$$

❏❏ **$\frac{40 rev}{min}$ equals how many rad/s?**

$$\frac{40 rev}{min}\left(\frac{.1047 rad/s}{1 rev/min}\right) \cong \frac{4.19 rad}{s}$$

❏❏ **Find the angular velocity of the second hand on a clock.**

$$\theta = wt \Rightarrow \frac{2\pi rad}{60 s} = w = .1047 \frac{rad}{s}$$

❏❏ **A turbine rotating at 5,000 rev/min comes to rest in 5 minutes. Find the angular acceleration that acts on the turbine.**

$$w = w_0 + \alpha t \Rightarrow \alpha = \frac{w - w_0}{t}$$

w_0 = 5,000 rev/min ≅ 523 rad/s

Thus, $\alpha = \frac{0 - 523 rad/s}{300 s} = -1.74 \frac{rad}{s^2}$

❏❏ **The earth orbits the sun once every 365 days. Find the angular velocity of the earth in its orbit.**

$$\theta = wt \Rightarrow \frac{2\pi rad}{(365 days)\left(\frac{86,400 s}{day}\right)} = w = 1.99 \times 10^{-7} \frac{rad}{s}$$

❏❏ **An object is executing uniform circular motion at 100 rev/min. (Note that 1 rev/min = 0.1047rad/s.) If the object is 5 cm from the axis of rotation, find the magnitude of the radial acceleration that acts on the object.**

$a_{rad} = rw^2 = (.05)(10.5 rad/s)^2 = 5.48 m/s^2$

☐☐ An electric motor runs at 1500 rev/min. Find (a) its angular speed and, (b) the linear speed of a point 6 cm from the axis of rotation.

(a) $1500 rev/min \left(\dfrac{.1047 rad/s}{1 rev/min} \right) = 157 rad/s$

(b) $v = rw \Rightarrow v = (.06m)(157 rad/s)$
$v = 9.42 m/s$

☐☐ A circular saw (r = 0.3m) starts from rest and accelerates with constant angular acceleration to a velocity of 200 rad/s in 15s. Find (a) the average angular velocity, (b) the angular acceleration, and (c) the angular displacement of the saw at the end of 15 s.

(a) $w_{av} = \dfrac{w + w_0}{2} = \dfrac{0 + 200 rad/s}{2} = 100 rad/s$

(b) $w = w_0 + \alpha t \Rightarrow \dfrac{w - w_0}{t} = \alpha = \dfrac{200 rad/s - 0}{15s}$
$\alpha = 13.3 rad/s^2$

(c) $\theta = w_{av} t = (100 rad/s)(15s) = 1500$ rad
or $\theta = (1/2)\alpha t^2 + w_0 t = (1/2)(13.3 rad/s)(15s)^2 + 0$
$\theta = 1500 rad$

☐☐ A wheel rotates with a constant angular velocity of 15 rad/s. Find the magnitude of the radial acceleration at a point 0.5m from the axis of rotation by (a) using $a_{radial} = v^2/r$, and (b) using $a_{radial} = rw^2$.

(a) $a_{radial} = \dfrac{v^2}{r}$ $v = rw = .5(15 rad/s)$
$v = 7.5$ m/s

$a_{radial} = \dfrac{(7.5 m/s)^2}{.5m} \approx 113 m/s^2$

(b) $a_{radial} = rw^2 = .5m(15 rad/s)^2 \approx 113$ m/s^2

☐☐ Find the required angular velocity in rev/min for an ultracentrifuge to have a radial acceleration of 300,000g [300,000 times the acceleration due to gravity] for a point 2 cm from the axis of rotation.

$a = rw^2 \Rightarrow \dfrac{(300,000)(9.8 m/s^2)}{.02m} = w^2 \Rightarrow w = 12,120 rad/s$

$w = 12,120 rad/s \left(\dfrac{1 rev}{2\pi rad} \right) \left(\dfrac{60s}{1 min} \right)$

$w \approx 115,800$ rev/min

◻◻ A car engine is idling at 600 rev/min. The accelerator is depressed and the engine speed increases to 4000 rev/min in 6 s. Find (a) the initial and final velocities in rad/s, (b) the angular acceleration that acts on the engine, (c) the angular displacement of the engine over the 6 s interval, (d) the linear velocity of a point on the flywheel that is 0.5m from the axis of rotation at the end of the 6 s interval and (e) the radial acceleration that acts on a point on the flywheel that is 0.4 m from the axis of rotation at the end of the 6 s interval.

(a) $w_0 = \dfrac{600 rev}{min}\left(\dfrac{.1047 rad/s}{1 rev/min}\right) = 62.8 rad/s$

$w = \dfrac{4000 rev}{min}\left(\dfrac{.1047 rad/s}{1 rev/min}\right) = 419 rad/s$

(b) $w = w_0 + \alpha t \Rightarrow \dfrac{w - w_0}{t} = \alpha = \dfrac{(419 rad/s - 62.8 rad/s)}{6s}$

$\alpha = 59.4$ rad/s

(c) $\theta = w_{av} t = \left(\dfrac{62.8 rad/s + 419 rad/s}{2}\right) 6s = 1445 rad$

(d) $v = rw \Rightarrow v = (.5m)(419 rad/s) = 210$ m/s

(e) $a_{radial} = rw^2 = (.4m)(419 rad/s)^2 = 70,200$ rad/s^2

◻◻ An electric motor increases its speed from 130 rad/s to 170 rad/s in 13 revolutions. Find (a) the angular acceleration that acts on the motor and (b) how long the acceleration acts on the motor.

(a) 1 rev = 2π rad $\Rightarrow$ 13 rev = 26π rad

$w^2 = w_0^2 + 2\alpha\theta$

$\dfrac{w^2 - w_0^2}{2\theta} = \alpha = \dfrac{(170 rad/s)^2 - (130 rad/s)^2}{2(26\pi rad)}$

$\alpha = 73.5$ rad/s

(b) $w = w_0 + \alpha t \Rightarrow t = \dfrac{w - w_0}{\alpha} = \dfrac{170 rad/s - 130 rad/s}{73.5 rad/s}$

t = .544s

◻◻ A rod of length 1 m is rotating at a constant 12 rad/s. What is the radial acceleration that acts at a point on the end of the rod?

$a_{radial} = rw^2 = (1m)(12 rad/s)^2 = 144$ m/s^2

☐☐ Suppose that the atmosphere was perfectly still. What would be the apparent wind velocity of a person on the equator? (The radius of the earth is about 6400 km.)

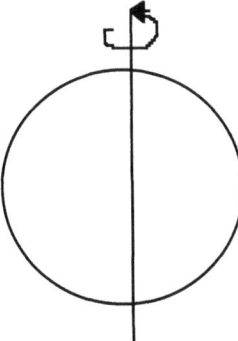

Here, the atmosphere is still, and the earth is rotating about its axis.
Thus, $\theta = w_{av} t$ $v = rw$
    2π rad $= w(86,400s)$ $= (6400 \times 10^3) 7.27 \times 10^{-5}$ rad/s
    $w = 7.27 \times 10^{-5}$ rad/s $v \cong 465$ m/s

☐☐ For the previous problem, what would be the wind velocity in mi/hr? (1mile = 1609m)

$v = 465 m/s \left(\frac{1 mile}{1609 m}\right)\left(\frac{3600 s}{1 hr}\right) = 1040 mi/hr$

☐☐ A bicycle has wheels of radius 0.38m. What is the angular displacement of a wheel on the bicycle after the bicycle has traveled a linear distance of 1 mile (1609m)?

$s = R\theta \Rightarrow 1609m = (.38m)\theta$
                $\theta = 4230$ rad

☐☐ After 8s, a wheel has slowed to an angular speed of 1.88 rad/s. During the time in which the wheel is slowing, the wheel rotates through an angular displacement of 50 rad. What angular acceleration acts on the wheel?

$\theta = w_{av} t \Rightarrow w_{av} = \frac{50 rad}{8s} = 6.25$ rad/s

Thus, $w_{av} = \frac{w + w_0}{2} \Rightarrow 2(6.25 rad/s) = 1.88 rad/s + w_0$
    $w_0 = 10.6$ rad/s

finally, $\theta = (1/2)\alpha t^2 + w_0 t \Rightarrow 50$ rad $= (1/2)(\alpha)(8s)^2 + (10.6 rad/s)8s$
    $\alpha = -1.09$ rad/s^2

☐☐ A rotor experiences a constant acceleration of 2.5 rad/s² and rotates through an angle of 290 rad in 10s. (a) What was the initial velocity of the rotor? (b) If the rotor started from rest, how long did it take to reach the velocity found in (a)?

(a) $\theta = \frac{1}{2}\alpha t^2 + w_0 t$

    290 rad $= .5(2.5$ rad/s$^2)(10s)^2 + (w_0)(10s)$

    w_0 for 10s interval $= 16.5$ rad/s

(b) to reach the beginning of the 10s interval

$$w = w_0 + \alpha t \Rightarrow \frac{w - w_0}{\alpha} = t = \frac{16.5 rad/s}{2.5 rad/s^2} = 6.6s$$

☐☐ **A fish is being reeled in at a speed of .15 m/s. The line is being spooled onto a reel whose radius is .03m. What is the angular speed of the reel?**

$v = rw \Rightarrow .15 m/s = (.03m)w$
$w = 5$ rad/s

☐☐ **A pump starts from rest and has an angular acceleration of 4 rad/s² for 16s. At the end of this time, find the (a) angular speed of the pump, and (b) the angle through which the pump has turned three different ways.**

(a) $w = w_{av} + \alpha t = 0 + (4 rad/s^2)(16s) = 64$ rad/s

(b) i. $\theta = w_{av} t = \left(\frac{0 - 64 rad/s}{2} \right) 16s = 512 rad$

ii. $\theta (1/2)\alpha t^2 + w_0 t = (1/2)(4 rad/s^2)(16s)^2 + 0 = 512$ rad

iii. $w^2 = w_0^2 + 2\alpha\theta \Rightarrow (64 rad/s)^2 = 0 + 2(4 rad/s^2)\theta$
$\theta = 512$ rad

ROTATIONAL MOTION II

THE FUNDAMENTAL IDEAS HERE ARE THE FOLLOWING:

- The moment of inertia I of an object that rotates is given by $I = m_1 r_1^2 + m_2 r_2^2 + m_3 r_3^2 + ...$ or $I = \Sigma m_i r_i^2$. The units of inertia are kg-m^2.
- An object that is rotating has kinetic energy. The kinetic energy of rotation is $K_{raot} = (1/2)Iw^2$. The units of rotational kinetic energy are Joules.
- It is torque that causes rotation. Torque τ is given by Force times lever arm (or moment arm). Thus $\tau = F(l.a.)$. Lever arm is perpendicular distance from the line of action of the Force to the axis of rotation. Note that torque can either be clockwise or counterclockwise.
  $\tau = 0$ if $F = 0$ or if $l.a. = 0$.
- Newton's Three Laws for rotational motion are:
  First Law: An object at rest remains at rest, an object in motion remains in motion
  unless acted upon by some external torque.
  Second Law: $\tau_{un} = I\alpha = \tau_{go} - \tau_{opp}$
  Third Law: For every torque there is an equal but opposite reaction (about the same axis).
- If an object is in a condition of equilibrium, it is at rest or moving with constant velocity. This means that
  1) $\Sigma F_{up} = \Sigma F_{down}$
  2) $\Sigma F_{left} = \Sigma F_{RT}$
  3) $\Sigma \tau_{cw} = \Sigma \tau_{ccw}$

- $\Sigma \tau_{cw} = \Sigma \tau_{ccw}$ is called the torque condition of equilibrium. The torque condition of equilibrium is meaningless until the axis of rotation is specified. In theory, if an object is in a condition of equilibrium, the axis of rotation may be place anywhere. In practice, it is placed at any convenient point or at that point about which rotation would take place.
- A table of analogs between linear and rotational motion is convenient.

Linear	Rotational
x(m)	θ(rad)
v(m/s)	w(rad/s)
a(m/s^2)	α(rad/s^2)
F(N)	τ(N-m)
m(kg)	I(kg-m^2)

- Thus, in linear motion $p = mv$ is linear momentum. Using the table, angular momentum $L = Iw$
- In linear motion, work = Fr (If $\theta = 0°$). In rotational motion, work = $\tau\theta$.

- Hence, if one knows linear motion, by using the table one can find the rotational equivalent, e.g,
  $$Power = \frac{work\ done}{time}$$
  $$P_{av} = \frac{Fr}{\tau} = Fv \Rightarrow P_{av} = \tau w$$

- From Newton's Second Law for rotational motion $\tau_{un} = I\alpha = \tau_{go} - \tau_{opp}$

- $\tau_{un} = \frac{I(w - w_0)}{\Delta t} \Rightarrow (\tau_{un})\Delta t = Iw - Iw_0$, that is, angular impulse = change in angular momentum.

- If $\tau_{ext} (= \tau_{un}) = 0 \Rightarrow Iw = Iw_0$ or $L = L_0$ which says that angular momentum is conserved during a collision.

SOLVED PROBLEMS

☐☐ An object (moment of inertia = 15 kg-m²) is rotating at 3,000 rev/min. What is its kinetic energy?

(note that 1 rev/min = .1047 rad/s.)

$$k = \frac{1}{2}Iw^2 = \frac{1}{2}(15kg - m^2)\left[3000\frac{rev}{min}\left(\frac{.1047 rad/s}{1 rev/min}\right)\right]^2$$
$$k = 7.4 \times 10^5 J$$

☐☐ For the object in the previous problem, how much work would the object do in coming to rest?

ΔE = work done $\Rightarrow$ work done = 7.4 x 10⁵ Joules.

☐☐ Consider the adjacent figure: (a) what is the torque about the axis caused by F_1? (b) What is the torque about the axis caused by F_2? (c) What is the resultant torque? (d) If the moment of inertia of the rod about the axis is 0.8 kg-m², what is the angular acceleration of the rod?

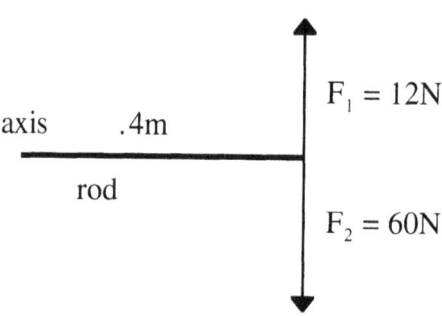

(a) $\tau = F(l.a.) = (12N)(.4m) = 4.8$ N-m ccw
(b) $\tau = F(l.a.) = (60N).4m = 24$ N-m cw
(c) $\tau_{resultant} = \tau_{cw} - \tau_{ccw} = 19.2$ N-m cw
(d) $\tau_{un} = I\alpha \Rightarrow (19.2 \text{N-m}) = (.8 kg\text{-}m^2)\alpha = 24$ rad/s²

GENERAL PHYSICS PEARLS OF WISDOM

❏❏ Consider the figure:

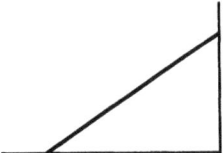

A uniform 5m ladder weighs 400N and rests against a wall with its lower end 3m from the wall. Find the horizontal and vertical components of the force that the ground exerts on the ladder and the force that the wall exerts on the upper end of the ladder. (Wall assumed to be friction-free)

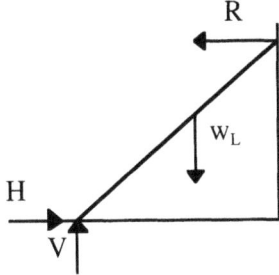

Since the ladder is in a condition of equilibrium, $\Sigma F_{up} = \Sigma F_{down}$

$v = w_L = 400N$

$\Sigma F_{left} = \Sigma F_{Right}$

$R = H$

$\Sigma \tau_{cw} = \Sigma \tau_{ccw}$.

By placing axis at foot of ladder,

$(w_L)l.a. = R(l.a.)$ $(l.a.)_{wL} = 1.5m$

$(400N)(1.5m) = R(4m)$ $(l.a.)_R = 4m$

$150N = R = H$

❏❏ A uniform meter stick is hinged as in the figure. It is nudged slightly. Find the angular velocity of the meter stick when it reaches the horizontal position. Assume no friction. (Note that $I = (1/3)ml^2$ for a rod hinged at one end)

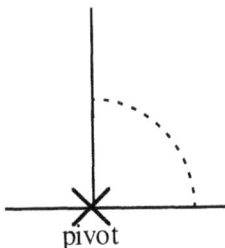

$E_0 = E \Rightarrow$ if the 0 of U_g is placed at the horizontal position, $U_g = k_{rot}$.

Thus $mg\dfrac{l}{2} = \dfrac{1}{2}Iw^2 = \dfrac{1}{2}\left(\dfrac{1}{3}ml^2\right)w^2$

so that $\dfrac{3g}{l} = w^2 = \dfrac{3(9.8m/s^2)}{1m} \Rightarrow w = 5.42 rad/s$

☐☐ A uniform plank (weight = 400N) is 10m long and supported at its ends by two walls. A person who weighs 800N is 3m from one end. By placing the axis of rotation for the torque condition at the left end, find the forces that the walls exert on the plank.

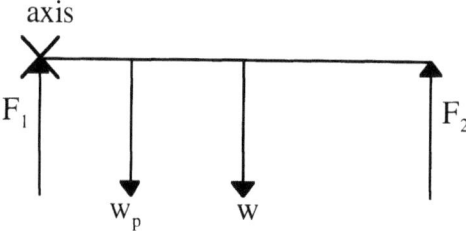

$\Sigma F_{up} = \Sigma F_{down}$

$F_1 + F_2 = w_p + w = 1200N$

$\Sigma \tau_{cw} = \Sigma \tau_{ccw}$

$(w_p)l.a. + (w)(l.a.) = (F_2)l.a. \Rightarrow (3m)800N + 400(5m) = F_2(10m)$

$F_2 = 440N$ and $F_1 = 760N$

☐☐ For the previous problem, place the axis of rotation at the right end of the plank and find the forces that the wall exert on the ends of the plank.

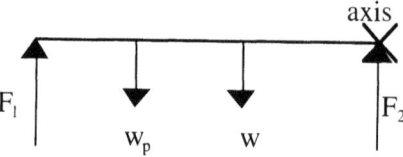

Again, $\Sigma F_{up} = \Sigma F_{down}$

$w_p + w = 400N + 800N = 1200N$ so that $F_1 + F_2 = 1200N$

$\Sigma \tau_{cw} = \Sigma \tau_{ccw}$

$(F_1)(l.a.) = (w_p)l.a. + (w)(l.a.)$

$(F_1)10m - (800N)(7m) + (400N)(5m)$

$F_1 = 760N \Rightarrow F_2 = 440N$ as before.

⧠⧠ An experimental jet turbine (moment of inertia is 1250kg-m²) is accelerated from rest to 15,000 rev/min in 40 seconds. Find (a) the torque needed to bring the turbine to its rated speed, (b) the work done on the turbine to bring it to its rated speed, (c) the kinetic energy of the turbine at its rated speed, (d) the size of the engine (in horsepower) required to bring the turbine to its rated speed, and (e) if an engine of 30,000 HP rating were used to bring the turbine to its rated speed, how long would it take?

(a) $\tau_{un} = I\alpha$ $\alpha = \dfrac{w - w_0}{t}$

$w_0 = 0$, $w = 15000 \dfrac{rev}{min} \left(\dfrac{.1047 rad/s}{1 rev/min} \right)$

≅ 1571 rad/s
⇒ α ≅ 39.3 rad/s²
so that τ_{un} = (1250kg-m²)39.3 rad/s² = 49,100N-m

(b) work done = ΔE
 i. Thus, work done = $(1/2)Iw^2$
    = $(1/2)(1250kg\text{-}m^2)(1571 rad/s)^2$
    = 1.54 x 10⁹J
 ii. or work = Fr in linear motion, so that, work = $\tau\theta = \tau(w_{av})t$
    = $(49100 N - m)\left(\dfrac{0 + 1571 rad/s}{2}\right) 40s$

work = 1.54 x 10⁹J

(c) $k_{rot} = (1/2)Iw^2$ = 1.54 x 10⁹J from above.

(d) $P_{av} = \dfrac{\Delta E}{time} = \dfrac{1.54 x 10^9 J}{40s} \left(\dfrac{1 H.P.}{746 J/s}\right) = 51,600 H.P.$

(e) $P_{av} = \dfrac{\Delta E}{time}$ ⇒ $time = \dfrac{\Delta E}{P_{av}} = \dfrac{1.54 x 10^9 J}{(30,000 H.P.)\left(\dfrac{746 J/s}{1 H.P.}\right)}$

time = 68.8s

⧠⧠ A 65kg grindstone is 0.8m in diameter and has a moment of inertia of 4kg-m². A tool is pressed against the grindstone with a normal force of 50N. If the coefficient of kinetic friction between the tool and the stone is 0.7, and if there is a constant friction torque of 5N-m in the bearings of the grindstone, (a) how much force must be applied (perpendicularly) to the end of a crank handle 0.6m long to bring the grindstone from rest to 150 rev/min in 10s, and (b) how long would it take the grindstone to come to rest if the force in (a) were removed after the grindstone achieved 150 rev/min? (Assume tool is also removed from grindstone.)

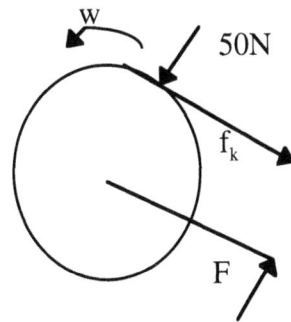

(a) $w_0 = 0$, $w = \dfrac{150 rev}{min}\left(\dfrac{.1047 rad/s}{1 rev/min}\right) = 15.7$ rad/s

$\alpha = \dfrac{w - w_0}{t} = 1.57 rad/s^2$

$\tau_{un} = I\alpha = \tau_{go} - \tau_{opp} = \tau_{go} - 5\text{N-m} - (f_k)\text{l.a.}$
$(1.57 rad/s^2)4kg\text{-}m^2 = (F_{go}\text{l.a.} - 5N - m - 35N(.4m))$
$F_{go} = 42.1N$

(b) $\tau_{un} = I\alpha = \tau_{goZ} - \tau_{opp}$
$4kg\text{-}m^2(\alpha) = 0 - 5\text{N-m}$
$\alpha = -1.25 rad/s^2$
Thus, $\dfrac{w - w_0}{t} = \alpha \Rightarrow \dfrac{-15.7 rad/s}{-1.25 rad/s} = t \cong 12.6s$

☐☐ Consider the boom in the figure. The boom is uniform and weighs 2000N, while the load weighs 6,000N. If the load is being lifted at constant speed, find (a) the tension in the guy wire (assume horizontal) and (b) the horizontal and vertical components of the force that acts on the end of the boom at the pivot.

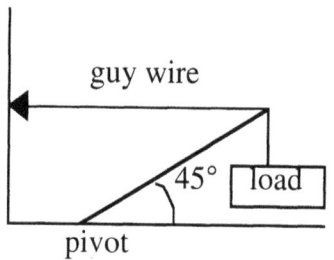

(a) placing the axis of rotation at the pivot, $\Sigma\tau_{cw} = \Sigma\tau_{ccw}$
$(w_b)(\text{l.a.}) + w_L(\text{l.a.}) = (T)\text{l.a.}$
assuming that the length of the boom is l, then
$(2000N)(l/2)\cos 45° + (6000N)l\cos 45° = Tl(\sin 45°)$
$\Rightarrow 7000N = T$

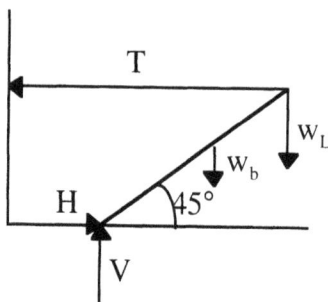

(b) $\Sigma F_{up} = \Sigma F_{down}$
$\Rightarrow V = w_b + w_L = 8000N$
$\Sigma F_{left} = \Sigma F_{right}$
$\Rightarrow T = H = 7000N$

❑❑ Consider the figure. If the string is pulled with force F, will the spool unwind or wind up?

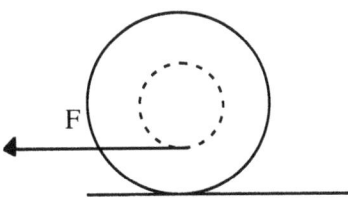

Wind up. The instantaneous axis of rotation is where the spool touches the table.

❑❑ Consider the box in the figure. The box is uniform and has a mass of 30 kg. The coefficient of static friction is 0.8. If F = 200N, what is the height h so that the box begins to tip?

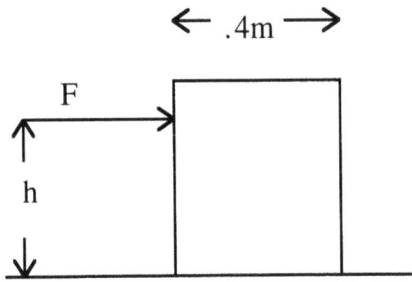

If the box just begins to tip, the axis of rotation is at the edge. Thus,
$\Sigma\tau_{cw} = \Sigma\tau_{ccw}$

(F)h = w(l.a.)
(200N)h = (25kg)(9.8m/s²)(.2m)
⇒ h = .245m

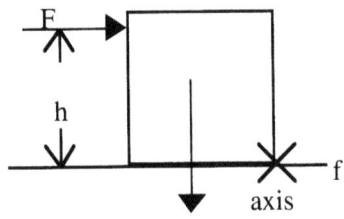

❑❑ What is the torque produced by an engine that develops 2000H.P. when it is rotating at 3000 rev/min?

P = Fv for linear motion ⇒ P = τw

Thus, $\tau = \dfrac{P}{w} = \dfrac{(2000 H.P.)\left(\dfrac{746 J/s}{1 H.P.}\right)}{(3000 rev/min)\left(\dfrac{.1047 rad/s}{1 rev/min}\right)} = 4750 N-m$

☐☐ The flywheel of an engine has a moment of inertia of 25kg-m². Find (a) the energy that the flywheel has when it is rotating at 4000 rev/min and (b) the work done in bringing the flywheel from rest to 4000 rev/min.

(a) $k = (1/2)Iw^2 = (1/2)(25kg\text{-}m^2)\left[4000\dfrac{rev}{min}\left(\dfrac{.1047 rad/s}{1 rev/min}\right)\right]^2 = 2.19 \times 10^6 J$

(b) ΔE = work done = $2.19 \times 10^6 J$

☐☐ A merry-go-round of radius 4m (moment of inertia is 15kg-m²) is rotating at 20 rad/s. A child (mass is 20kg) jumps on the outer edge of the merry-go-round. What is the angular velocity of the merry-go-round after the person has jumped on? (Ignore friction).

Here $\tau_{ext} = 0 \Rightarrow (Iw)_0 = Iw$
Thus, since $I_0 = 15kg\text{-}m^2$
$I = 15kg\text{-}m^2 + m_p r_p^2$
$= 15kg\text{-}m^2 + (20kg)(16m^2)$
$= 335kg\text{-}m^2$

Thus, $(15kg\text{-}m^2)20 rad/s = (335kg\text{-}m^2)w$
$w = .895 rad/s$

☐☐ An object experiences an angular acceleration of 12rad/s² due to the application of a torque of 15N-m. What is the moment of inertia of the object, assuming there is no friction?

$\tau_{un} = I\alpha = \tau_{go} - \tau_{opp}$

$\Rightarrow I = \dfrac{\tau_{go}}{\alpha} = \dfrac{15 N-m}{12 rad/s^2} = 1.25 \dfrac{(kg-m/s^2)m}{rad/s^2} = 1.25 kg-m^2$

☐☐ A uniform ladder that is 8m long and weighs 350N has its lower end 3m from a smooth vertical wall. A person who weighs 800N is 5m up the ladder. Find (a) the force that the wall exerts on the upper end of the ladder, and (b) the horizontal and vertical components of the force that the ground exerts on the base of the ladder.

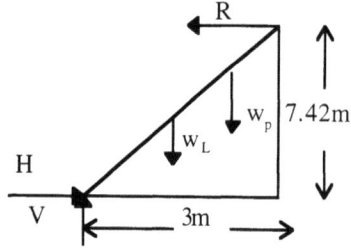

(a) Placing the axis of rotation at the bottom of the ladder, $\Sigma\tau_{cw} = \Sigma\tau_{ccw}$
   $(w_L)l.a. + (w_p)l.a. = (R)l.a.$
   $(350N(1/2)3m + (800N)(5/8)3m = R(7.42m)$
   $R = 273N$

(b) $\Sigma F_{up} = \Sigma F_{down}$
   $V = w_L + w_p = 1150N$
   $\Sigma F_{left} = \Sigma F_{right}$
   $\Rightarrow R = H = 273N$

❏❏ For the previous problem, why did placing the axis of rotation at the bottom of the ladder eliminate H and V from the torque condition of equilibrium?

Since $\tau = (F)(l.a.)$, placing the axis of rotation at the bottom eliminates H and V from the torque condition of equilibrium because their lever arms are zero. Hence they contribute no torque.

❏❏ A uniform plank of weight 450N and length 8m is supported by two supports that are 2m from each end. Could a person who weighs 950N walk to the end of the plank without causing it to tip?

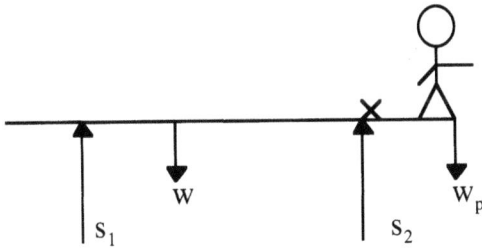

The axis of rotation would be about S_2 if the plank was going to tip. Thus,
$\tau_{cw} = w_p l.a. = (950N)2m = 1900N\text{-m}$
$\tau_{ccw} = w l.a. = (450N)4m = 1800N\text{-m}$
Since the clockwise torque is greater than the counter clockwise torque, the plank would tip.

❏❏ For the previous problem, how far from support s_2 could the person walk before the plank began to tip?

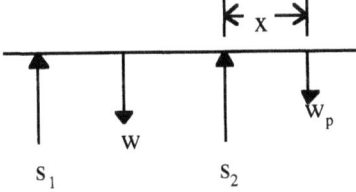

$\Sigma\tau_{cw} = \Sigma\tau_{ccw}$

$(w_p)x = (w)l.a.$

$(950N)x = (450N)(4m)$

$x = \dfrac{1800N\text{-}m}{950N} = 1.89m$ from support or person would be .11m from end before the plank started to tip.

❏❏ A flywheel 1.2m in diameter is pivoted about an axis through its center and is at rest. A rope is wrapped around the flywheel, and a constant pull of 50N is applied to the flywheel for 5s, which causes 8m of rope to be unwound. (a) What is the angular acceleration of the flywheel? (b) What is the final angular velocity of the flywheel? (c) What is the final kinetic energy of the flywheel? (d) What is the moment of inertia of the flywheel? (Ignore friction.)

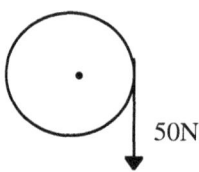

(a) $a_T = \alpha r$ $x = (1/2)at^2 + v_0 t$
$\Rightarrow 8m = (1/2)(a)(25s^2)$
$a = .64 m/s^2$
Thus, $.64 \, m/s^2 = \alpha(.6m)$
$\Rightarrow \alpha = 1.07 \, rad/s^2$

(b) $w = w_0 + \alpha t$
$w = 0 + (1.07 rad/s^2)(5s) \Rightarrow w = 5.35 \, rad/s$

(c) $\Delta E = $ work done $\Rightarrow k = (50N)(8m) = 400 N\text{-}m$

(d) $k = (1/2)Iw^2 \Rightarrow \dfrac{2k}{w^2} = I = \dfrac{800 N - m}{(5.35 rad/s)^2} = 28 kg - m^2$

⊡⊡ **A merry-go-round is rotating at 6.28 rad/s, and has a moment of inertia of 150kg-m². A child (mass is 50kg) jumps on the merry-go-round 1.6m from the center. Find the resulting angular speed of the merry-go-round. (Ignore friction.)**

$Iw_0 = Iw$ because $\tau_{ext} = 0$.
Thus, $(150 kg\text{-}m^2)(6.28 rad/s) = [150 kg\text{-}m^2 + 50 kg(1.6m)^2]w$
$w = 3.39 \, rad/s$

⊡⊡ **A uniform door (.8m wide, 2.2m high) weighs 150N and is supported by two hinges that are set in .1m from the top and the bottom of the door. Find the direction and magnitude of the horizontal component of the force that each hinge exerts on the door.**

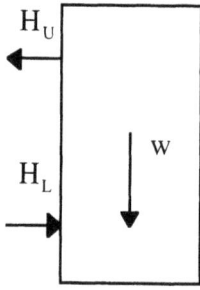

By using a book, it is easy to determine that the direction of the horizontal component that each hinge exerts on the door is as in the figure.
$\Sigma F_{left} = \Sigma F_{right} \Rightarrow H_U = H_L$
By placing the axis of rotation at the lower hinge
$\Sigma \tau_{cw} = \Sigma \tau_{ccw} = (w)l.a. = (H_U)l.a.$
$= (150N)(.4m) = (H_U)2m$
$H_U = H_L = 30N$

☐☐ An electric motor is rotating at 6 rad/s and is shut off, coming to rest in 16s. If the moment of inertia of the motor is 1.2kg-m², (a) what angular acceleration acts on motor while it is coming to rest, (b) what torque acts on the motor while it is coming to rest, and (c) what work was done on the motor to bring it to rest?

(a) $\alpha = \dfrac{w - w_0}{t} = \dfrac{0 - 6 rad/s}{16s} = -.375 rad/s$

(b) Thus, $\tau_{opp} = (1.2\text{kg-m}^2)(-.375\text{rad/s}) = -.45\text{N-m}$

(c) method 1: work $= \tau\theta = \tau w_{av} t$
    work $= (-.45\text{N-m})(3\text{rad/s})(16\text{s})$
    work $= -21.6\text{J}$

method 2: ΔE = work done
   $\Rightarrow$ work done $= E - E_0 = 0 - (1/2)I w_0^2$
   $= (-1/2)[1.2\text{kg-m}^2](6\text{rad/s})^2$
   work done $= -21.6\text{J}$

☐☐ A revolving door consists of four rectangular doors (each of moment of inertia 50kg-m²) that rotate about a central axis. If a person pushes on the outer edge of one of the doors with a force of 80N (directed perpendicularly to the door), what angular acceleration acts on the door if friction is ignored? (Assume door starts from rest.)

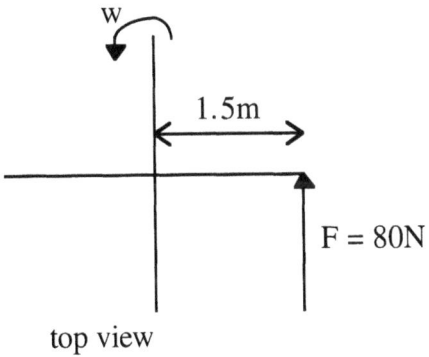

top view

$\tau_{un} = I\alpha = \tau_{go} - \tau_{opp}$

$(4)(50\text{kg-m}^2)\alpha = 80\text{N}(1.5\text{m})$

$\alpha = .6 \text{ rad/s}^2$

❏❏ **A pair of forces that have equal magnitude but opposite directions (and different lines of action) is called a couple. Show that the torque produced by the couple in the adjacent figure is independent of whether the axis is at point A, B, or C.**

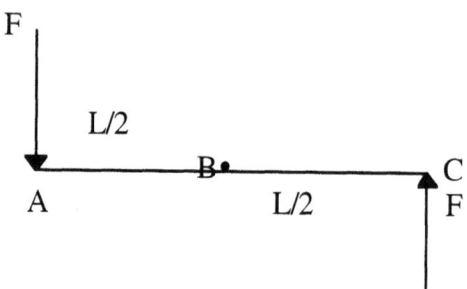

axis at point A:

$\tau = Fl.a. = Fl_{ccw}$

axis at point B

$\tau = F(L/2) + F(L/2) = Fl_{ccw}$

axis at point C

$\tau = F(L) = Fl_{ccw}$

ELASTICITY AND SIMPLE HARMONIC MOTION

THE FUNDAMENTAL IDEAS HERE ARE THE FOLLOWING:

- Stress equals force divided by area, F/A.

- Strain equals change in quantity divided by the quantity, (ΔQ)/Q.

- Modulus equals stress divided by strain.

- Young's modulus $y = \dfrac{stress}{strain} = \dfrac{F/A}{\Delta l / l}$

- Bulk (volume) modulus $B = \dfrac{F/A}{\Delta V / V} = -\dfrac{\Delta P}{\Delta V / V}$
  the negative is because an increase in pressure causes a decrease in volume.

- Compressibility k is the reciprocal of the bulk modulus B. Thus k = 1/B. The units are atm^{-1} or m^2/N.

- The elastic limit is defined to be maximum stress from which the object can regain its original size or shape.

- The ultimate tensile stress (or breaking stress) is defined to be the maximum stress the object can sustain.

- Simple harmonic motion (SHM) is motion in which the acceleration is proportional to the displacement but in the opposite sense.

- Some definitions from SHM are the following:
  a. equilibrium position → the displacement = 0 position
  b. amplitude A → the maximum displacement from the equilibrium position
  c. period T → the time for one round trip (or one revolution, cycle, oscillation)
  d. frequency f → number of round trips divided by time. The unit of frequency is cycles/s or Hertz(Hz).
  e. angular frequency w → 2πf

- for SHM, T = 1/f and $T = 2\pi \sqrt{\dfrac{-displacement}{acceleration}}$

- A reference circle is sometimes useful.

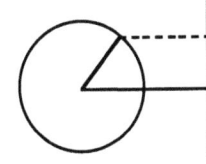

- An object that is executing uniform circular motion has a shadow that is executing SHM.

- For a mass on the end of a spring, $T = 2\pi \sqrt{\frac{m}{k}}$ and the mass executes SHM.

- For a mass on the end of a string, $T = 2\pi \sqrt{\frac{l}{g}}$, if the angular displacement from the equilibrium position is small (< 10°) and the mass executes SHM.

- A physical pendulum is an object oscillating about an axis with small angular displacements.

- Acceleration in SHM is given by $a = -4\pi^2 \left(\frac{y}{T^2}\right)$, where T is period and y is displacement from the equilibrium position.

- From Newton's 2nd Law, $F = ma = \frac{-4\pi^2 m y}{T^2}$

- Energy in SHM can be either elastic potential energy or kinetic energy or the sum of the $U_s + k$.

- Velocity in SHM is given by $v = \frac{2\pi}{T}\sqrt{A^2 - y^2}$, where A is amplitude, y is displacement, and T is period.

- $E = K + U = (1/2)mv^2 + (1/2)ky^2 = (1/2)kA^2$, thus,

$$v^2 = \frac{kA^2 - ky^2}{m}$$

$$\Rightarrow v = \pm\sqrt{\frac{k}{m}}\sqrt{A^2 - y^2}$$

$$= \pm w\sqrt{A^2 - y^2}$$

SOLVED PROBLEMS

☐☐ **A wire 2m long has a radius of 0.6mm. It is stretched 0.4mm when a 7kg mass is hung on the wire. Find Young's modulus for the material.**

$$Y = \frac{F/A}{\Delta l / l} = \frac{\frac{(7kg)(9.8m/s^2)}{\pi(6\times10^{-4m})^2}}{\frac{4\times10^{-4m}}{2m}} = 3.03\times10^{11} N/m^2$$

☐☐ **For the wire in the previous problem, find the elongation (how much the wire stretches) when a load of 500N is suspended at its end.**

$$Y = \frac{F/A}{\Delta l / l} \Rightarrow \frac{\Delta l}{l} = \frac{F}{Ay} \Rightarrow \Delta l = \frac{Fl}{Ay}$$

Thus, $\Delta l = \dfrac{(500N)(2m)}{\pi(6\times10^{-4}m)^2(3.03\times10^{11}N/m^2)} = 2.92\times10^{-3}m$

☐☐ A hydraulic jack contains 0.2 m³ of oil that has a compressibility of 2x10⁻⁵/atm. Calculate the decrease in volume if it is subjected to a pressure of 500 atm. Find the Bulk modulus of the oil in N/m².

$$B = \frac{-P}{\Delta v/v} \Rightarrow \frac{\Delta v}{v} = \frac{-P}{B}, \text{ thus, } \Delta v = -Pvk = (500 atm)(.2m^3)\left(\frac{2x10^{-5}}{atm}\right)$$

so that $\Delta v = 2x10^{-3} m^3$

The Bulk modulus can be found by using the definition of compressibility k:
$$k = \frac{1}{B} \Rightarrow B = \frac{1}{k} = \frac{1}{2x10^{-5}/atm} = 5x10^4 atm$$
$$/atm = 1.01x10^5 N/m^2 \Rightarrow B = 5.05x10^9 N/m^2$$

☐☐ A simple pendulum (length = 1.8m) makes 180 round trips in 67s. Find (a) the period, (b) the frequency, (c) the angular frequency, and (d) the acceleration due to gravity at the location of the pendulum.

(a) T = time for one round trip, thus $T = \frac{180 \text{ round trips}}{67s} = 2.69s$

(b) frequency f = 1/T = .377 cycles/s or .372 Hz

(c) w = 2πf = 2.34 rad/s

(d) $T = 2\pi \sqrt{\frac{l}{g}} \Rightarrow g = \frac{4\pi^2 l}{T^2} = \frac{4\pi^2(1.8m)}{(2.69s)^2} = 9.81 \frac{m}{s^2}$

☐☐ The object in the figure is executing SHM. It takes 2.4s to go from x = .6m to x = -.6m. Find (a) the period, (b) the frequency, (c) the total distance traveled for one period, (d) the angular frequency, (e) the maximum speed, (f) the velocity when x = 0.1m, (g) the acceleration when x = 0.4m and (h) the maximum acceleration when x = 0.2m.

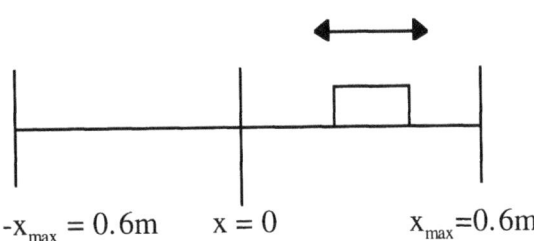

$-x_{max} = 0.6m$ $x = 0$ $x_{max} = 0.6m$

(a) T = 4.8s (time for one round trip)

(b) f = 1/T = 1/4.8s = .208Hz

(c) For one round trip, distance traveled = 2.4m

(d) w = 2πf = 1.31 rad/s

(e) $v = \frac{2\pi}{T}\sqrt{A^2 - x^2}$ this will be a maximum when $x = 0$

$\Rightarrow v_{max} = \frac{2\pi}{T} A = \frac{6.28}{4.8s}(.6m)$

$v_{max} = .785 m/s$

(f) $v_{x=0.1m} = \frac{2\pi}{T}\sqrt{A^2 - x^2} = \frac{6.28}{4.8s}\sqrt{(.6m)^2 - (.1m)^2}$

$v_{x=0.1m} = .774 m/s$

(g) $T = 2\pi \sqrt{\frac{-disp}{acc}} \Rightarrow a = \frac{-4\pi^2 disp}{T^2} \Rightarrow a_{x=.4m} = \frac{-4\pi^2(.4m)}{(4.8s)^2}$

Thus $a_{x=.4m} = -.685 m/s^2$

(h) $a_{max} = \frac{-4\pi^2 (disp)_{max}}{T^2} = \frac{-4\pi^2(.6m)}{(4.8s)^2} = -1.03 m/s^2$

☐☐ A 0.5kg mass on the end of a spring is executing SHM with a frequency of 1.2 Hz. Find (a) the period of the motion, (b) the angular frequency of the motion, and (c) the spring constant of the spring.

(a) $T = \frac{1}{f} = \frac{1}{1.2 Hz} = .833s$

(b) $w = 2\pi f = 2\pi(1.2 Hz) = 7.54 rad/s$

(c) For a mass on the end of a spring, $T = 2\pi \sqrt{\frac{m}{k}} \Rightarrow T^2 = 4\pi^2 \frac{m}{k}$ or $k = \frac{4\pi^2 m}{T^2}$

Thus, $k = \frac{4\pi^2(.5kg)}{(.833s)^2} = 28.4 kg/s^2$ or $28.4 N/m$

This problem could have been worked by using the frequency. Since

$f = \frac{1}{T} \Rightarrow f = \frac{1}{2\pi}\sqrt{\frac{k}{m}}$ or $f^2 4\pi^2 m = k = (1.2 Hz)^2 (4\pi^2)(.5kg)$

Thus, $k = 28.4 kg/s^2$ or $28.4 N/m$

☐☐ If the period of a simple pendulum is to be 2s, what should its length be if it is at a place where $g = 9.78 m/s^2$?

For a simple pendulum, $T = 2\pi \sqrt{\frac{l}{g}} \Rightarrow l = \frac{T^2 g}{4\pi^2} = \frac{(4s^2)9.78 m/s^2}{4\pi^2} = .992 m$

☐☐ A metal rod 5m long and 0.6cm² in cross section is found to stretch 0.15cm under a tension of 15,000N. For these conditions, find (a) the stress, (b) the strain, and (c) Young's modulus for the material.

(a) stress = $\dfrac{F}{A} = \dfrac{15,000N}{0.6cm^2} = 25,000 N/cm^2$

stress = $2.5 \times 10^8 N/m^2$

(b) strain = $\dfrac{\Delta l}{l} = \dfrac{0.15 \times 10^{-2} m}{5m} = 3 \times 10^{-4}$

(c) Young's modulus(Y) = $\dfrac{\frac{F}{A}}{\frac{\Delta l}{l}} = \dfrac{2.5 \times 10^8 N/m^2}{3 \times 10^{-4}}$

Y = $8.33 \times 10^{11} N/m^2 = 8.33 \times 10^{11} Pa$

☐☐ Suppose that a cubic meter of water is taken from the surface of the ocean to the bottom of the Marianas Trench, where the pressure is about 1090atm. What is the change in volume of the cubic meter of the water if the compressibility of water is 46.4×10^{-6}/atm?

Bulk modulus B = $\dfrac{-\Delta P}{\frac{\Delta V}{V}} \Rightarrow \dfrac{\Delta V}{V} = -\dfrac{\Delta P}{B}$

Thus, $\Delta V = \dfrac{-(\Delta P)V}{B} = -(\Delta P)Vk = -(1090 atm)(1m^3)\left(\dfrac{46.4 \times 10^{-6}}{atm}\right)$

so $\Delta V = -5.05 \times 10^{-2} m^3$

☐☐ A simple pendulum on earth has a period of 3s on earth. What would be its period on the moon, where the acceleration due to gravity is 1.67m/s²?

The period for a simple pendulum is given by T = $2\pi \sqrt{\dfrac{l}{g}}$

Thus, $T_m = 2\pi \sqrt{\dfrac{l}{g_m}}$ and $T_E = 2\pi \sqrt{\dfrac{l}{g_E}}$

Squaring each equation and dividing one by the other:

$\dfrac{T_m^2}{T_E^2} = \dfrac{4\pi^2 \left(\dfrac{l}{g_m}\right)}{4\pi^2 \left(\dfrac{l}{g_E}\right)} = \dfrac{g_E}{g_m}$ so that $T_m^2 = \dfrac{T_E^2 g_E}{g_m} = \dfrac{9s^2(9.8m/s^2)}{(1.67m/s^2)}$

$T_m^2 = 52.8 s^2$ so that $T_m = 7.27s$

◻◻ A small object is undergoing SHM in a horizontal plane with an amplitude of 8cm. At a point 5cm away from equilibrium, the velocity is 20cm/s. Find (a) the period of the motion, (b) the frequency of the motion, (c) the displacement when the velocity is ± 15cm/s, and (d) the maximum acceleration that acts on the object.

(a)
$$v = \frac{2\pi}{T}\sqrt{A^2 - x^2}$$
$$20 cm/s = \frac{6.28}{T}\left([8cm]^2 - [5cm]^2\right)^{1/2}$$
$$T = 1.96s$$

(b) $f = 1/T = .51 Hz$

(c) $v = \frac{2\pi}{T}\sqrt{A^2 - x^2}$

$$\pm 15 cm/s = \frac{6.28}{1.96s}\sqrt{(8cm)^2 - x^2}$$

$$\pm 4.68 = \sqrt{(8cm)^2 - x^2} \Rightarrow 21.9 cm^2 = 64 cm^2 - x^2$$
$$x = \pm 6.49 cm$$

(d) $T = 2\pi\sqrt{\frac{-disp}{acc}} \Rightarrow a = \frac{-4\pi^2}{T^2} disp$

$$a_{max} = \frac{-4\pi^2}{(1.96s)^2} 8cm = -82.1 cm/s^2 = -.821 m/s^2$$

◻◻ A small object oscillates back and forth at the bottom of a frictionless bowl in a hemispherical shape of radius R. Show that the motion is SHM, and find the period of the motion as a function of R and g. Assume θ is small.

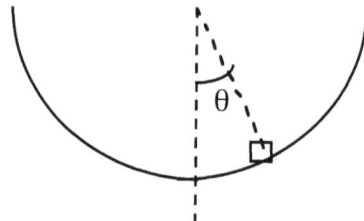

To find out whether or not the motion is SHM, apply Newton's 2nd Law for rotation.
$\tau_{un} = I\alpha = \tau_{go} - \tau_{opp}$

Once the object is set in motion: $I\alpha = 0 - mgsin\theta$
but
$\quad I = m_1 r_1^2 + m_2 r_2^2 + ...$
$\Rightarrow I = mR^2$

Thus, $\alpha = \frac{-mgR\sin\theta}{mR^2} = \frac{-g}{R}\theta \Rightarrow$ motion is SHM because acceleration is proportional to displacement but in opposite sense.

Since $T = 2\pi \sqrt{\dfrac{-Disp}{acc}} = 2\pi \sqrt{\dfrac{-\theta}{-\dfrac{g}{R}}}$ so that $T = 2\pi \sqrt{\dfrac{R}{g}}$

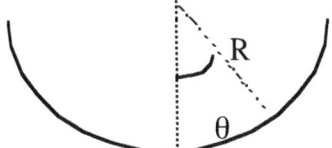

☐☐ **Given that the elastic limit stress for aluminum is 1.38x10⁸Pa, what is the maximum load that could be applied to an aluminum rod of cross section 0.5cm² without exceeding the elastic limit?**

Stress = F/A ⇒ F = (stress)A. 0.5cm² = 0.5x10⁻⁴ m². Then the force that would take the rod to the elastic limit is (1.38x10⁸N/m²)(0.5x10⁻⁴m²). Thus, the force that would take the rod to the elastic limit is 6900N.

☐☐ **A steel post that has a radius of 6cm is 4m long and supports a load of 12,000kg. Find (a) the stress in the post, (b) the strain in the post and (c) the change of length of the post. (Young's modulus for steel is 2x10¹¹N/m²)**

(a) stress = F/A = $\dfrac{(12,000 kg)(9.8 m/s^2)}{\pi(.06 m^2)}$ = $1.04 \times 10^7 \, N/m^2$

(b) strain = $\dfrac{\Delta l}{l}$. From young's modulus:

$Y = \dfrac{\dfrac{F}{A}}{\dfrac{\Delta l}{l}} \Rightarrow \dfrac{\Delta l}{l} = \dfrac{F}{Ay} = \dfrac{1.04 \times 10^7 \, N/m^2}{2 \times 10^{11} \, N/m^2}$

$\dfrac{\Delta l}{l} = 5.2 \times 10^{-5}$

Thus, the strain = 5.2x10⁻⁵

(c) $\dfrac{\Delta l}{l} = 5.2 \times 10^{-5} \Rightarrow \Delta l = (5.2 \times 10^{-5})(l)$

Δl = (5.2x10⁻⁵)(4m)
Δl = 2.08x10⁻⁴m

☐☐ **If the ultimate tensile stress (or breaking stress) for aluminum is 1.4x10⁸N/m², What is the maximum load that can be supported by an aluminum rod that is 1cm thick?**

Ultimate tensile stress = $F_{max}/A \Rightarrow A(U.T.S.) = F_{max}$

$\Rightarrow F_{max} = \left(\dfrac{\pi D^2}{4}\right) U.T.S. = \dfrac{(3.14)(.01m)^2}{4}(1.4 \times 10^8 \, N/m^2) = 11,000N$

☐☐ An object is hanging from a spring balance that has a scale that reads from 0 to 250N and is 12cm long. The object is executing SHM with a period of 1.5s. Find the mass of the object.

$F = kx \Rightarrow 250N = (.12m)k \Rightarrow k \approx 2080 N/m$

for a mass on the end of a spring, $T = 2\pi \sqrt{\frac{m}{k}} \Rightarrow T^2 = 4\pi^2 \frac{m}{k}$

so that $m = \frac{T^2 k}{4\pi^2} = \frac{(1.5s)^2 (2080 N/m)}{4\pi^2}$

Thus, the mass is 119kg.

☐☐ An object is undergoing SHM with an amplitude of 16cm and a frequency of 5Hz. Find (a) the maximum acceleration, (b) the minimum acceleration, (c) the maximum velocity, (d) the minimum velocity, and (e) the time required for the object to go from the equilibrium position to a point 14cm from it.

(a) $T = \frac{1}{f} = \frac{1}{5Hz} = 0.2s$

$T = 2\pi \sqrt{\frac{-Disp}{acc}} \Rightarrow a = \frac{-4\pi^2}{T^2} y$

$\Rightarrow a_{max} = \frac{-4\pi^2}{T^2} y_{max} = \frac{-4\pi^2}{(0.2s)^2} (.16m) = 158 m/s^2$

(b) from part (a) above $a_{min} = 0$ when $y = 0$

(c) $v = \frac{2\pi}{T} \sqrt{A^2 - y^2} \Rightarrow v_{max} = \left(\frac{2\pi}{T}\right) A$ when $y = 0$

$v_{max} = \left(\frac{6.28}{.2s}\right).16m = 5.02 m/s$

(d) $v_{min} = 0$ when $y = y_{max} = A$

(e) here, the reference circle is of help.

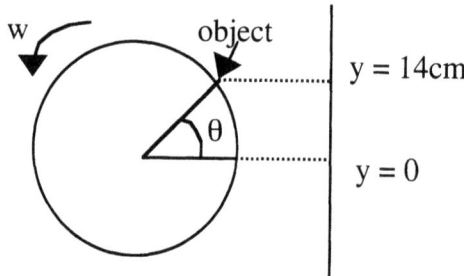

Assume that an object is undergoing uniform circular motion. It is easy to show that the object's shadow is undergoing SHM. Thus, the time it takes for the object to undergo an angular displacement θ is the time it takes for the shadow to move from the $y = 0$ position to the $y = 14m$ position. Thus, for the object, $\theta = wt \Rightarrow t = \frac{\theta}{w}$

for the shadow, $w = 2\pi f = 31.4$ rad/s

Hence, $t = \dfrac{1.07 \, rad}{31.4 \, rad/s} = .034s$

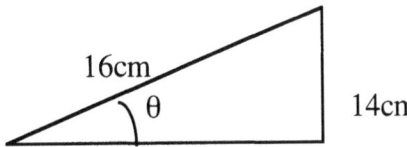

$\sin\theta = 14cm/16cm = .875 \Rightarrow \theta = 61°$
or $\theta \cong 1.07 \, rad$

☐☐ **A force of 40N stretches a spring 18cm. What mass must be attached to the spring so that the system will oscillate with a period of $(\pi/8)$s?**

For a mass on the end of a spring, $T = 2\pi\sqrt{\dfrac{m}{k}} \Rightarrow T^2 = \dfrac{4\pi^2 m}{k} \Rightarrow \dfrac{T^2 k}{4\pi^2} = m$

Thus, from Hooke's Law, $F = kx$

$40N = (.18m)k \Rightarrow k = 222N/m$. Hence, $\dfrac{(\pi^2/64)s^2}{4\pi^2}(222N/m) = m = 0.867kg$

☐☐ **A small block is executing SHM with an amplitude of 12cm. At a point 6cm from the equilibrium position, the velocity is 24cm/s. Find (a) the period, (b) the frequency, and (c) the displacement when the velocity is ± 15cm/s**

(a) $v = \dfrac{2\pi}{T}\sqrt{A^2 - x^2} \Rightarrow 24cm/s = \dfrac{6.28}{T}\sqrt{(12cm)^2 - (6cm)^2}$.

Thus, the period is 2.72s

(b) $f = \dfrac{1}{T} = .367Hz$

(c) $v = \dfrac{2\pi}{T}\sqrt{A^2 - x^2} \Rightarrow \pm15cm/s = \dfrac{6.28}{2.72s}\sqrt{(12cm)^2 - x^2}$

$\Rightarrow x = \pm 10.1cm$ when $v = \pm 15cm$

FLUIDS

THE FUNDAMENTAL IDEAS ARE THE FOLLOWING:

- Mass density = $\dfrac{mass}{volume}$, $\rho = \dfrac{m}{V}$

- Thus, weight w = mg = ρvg

- Pressure P = $\dfrac{F}{A}$ $(F \perp A)$ or $\Delta P = \dfrac{F}{A} (F \perp A)$

- The pressure due to a column of fluid is given by $P_{fluid} = \rho gh$, where h is the height of the fluid above the point in question.

- Gauge pressure is pressure in excess of atmospheric pressure.

- Absolute pressure is total pressure.

- Pascal's principle says that any change in pressure for an enclosed fluid at rest is transmitted undiminished throughout the fluid.

- Archimedes' Principle says that an object immersed in a fluid is buoyed (lifted) up by a force equal to the weight of the fluid displaced.

- An ideal fluid is a fluid that is incompressible, steady, irrational, and non-viscous (no friction between layers of the fluid).

- The path of an element of a moving fluid is called a flowline.

- A streamline is a curve whose tangent at any point is in the direction of the fluid velocity at that point.

- A flow tube is bounded by flowlines.

- For an ideal fluid the volume flow rate is constant.

- The volume flow rate for an ideal fluid can be written as $A_2v_2 = H_2v_2$ where A is the cross sectional area of the tube of flow and v is the velocity of the fluid at that point. This is called the continuity equation.

- It can be easily demonstrated for fluids in motion that where the velocity is greater the pressure is less.

- For an ideal fluid along a flow tube, Bernoulli's Theorem says that the pressure plus the kinetic energy per unit volume plus the potential energy per unit volume is a constant, i.e.
$p_1 + \dfrac{1}{2}\rho v_1^2 + \rho gh_1 = p_1 + \dfrac{1}{2}\rho v_2^2 + \rho gh_2$ where the pressure is absolute pressure and h is the height above the 0 of U_g.

SOLVED PROBLEMS

☐☐ **A force of 50N is applied over an area of 5mm². What is the pressure?**

$$P = \frac{F}{A} = \frac{50N}{5x10^{-6}m^2} = 1x10^7 N/m^2 Pa = 1x10^7 Pa$$

☐☐ **Sea water has a density of 1030kg/m³. What is the weight of 1m³ of seawater?**

w = mg, but ρ = M/v ⇒ m = ρv. Thus, w = ρvg = (1030kg/m³(1m³)9.8m/s²
w = 10,100N

☐☐ **What is the fluid pressure at the bottom of the Marianas Trench, which is 10,900m deep? Assume the density of seawater is 1030kg/m².**

$$\rho_{fluid} = \rho gh = (1030kg/m^3)(9.8m/s^2)(10,900m) = 1.1x10^8 Pa$$

☐☐ **For the previous problem, express the fluid pressure in atmospheres, given that 1 atm = 1.01x10⁵N/m².**

$$(1.1x10^8 N/m^2)\left(\frac{1atm}{1.01x10^5 N/m^2}\right) = 1090 atm$$

☐☐ **A car plunges into a lake and settles to the bottom. The driver can't get the door open. Explain why.**

ΔP = F/A ⇒ F = (ΔP)A
The pressure due to the fluid creates a pressure difference force holding the door shut. She needs to roll down or break a window to equalize the pressure so that the pressure difference force holding the door shut is reduced to zero.

☐☐ **An object weighs 16N. What is its mass density if its volume is 224cm³?**

$$\rho = \frac{m}{v} = \frac{16N/9.8m/s^2}{224x10^{-6}m^3} = 7290 kg/m^3$$

☐☐ **What is the specific gravity of the object in the previous problem? (ρ_{H20} = 1000kg/m³)**

$$S.G. = \frac{\rho_{obj}}{\rho_{H20}} = \frac{7290 kg/m^3}{1000 kg/m^3} = 7.29$$

☐☐ **A balloon has a volume of 700 m³. It is filled with helium (ρ = 0.178 kg/m³). If the density of air is 1.29 kg/m³, find (a) the buoyant force that acts on the balloon, (b) the weight of the helium used to inflate the balloon, and (a) the resultant force that acts on the balloon. (Assume volume of air displaced equals the volume of the helium in the balloon and ignore the weight of the balloon material and equipment.)**

(a) To find the buoyant force that acts on the balloon,
$$F_{buoy} = w_{air\ displaced} = \rho v g = (1.29 kg/m^3)(700 m^3)(9.8 m/s^2)$$
$$F_{buoy} = 8850\ N$$

(b) To find the weight of the helium in the balloon,
$$w_{helium} = \rho v g_{He} = (0.178 kg/m^3)(700 m^3)(9.8 m/s^2)$$
$$w_{helium} = 1220\ N$$

(c) The resultant force equals the force up minus the force down, thus,
R = 8850 N - 1220 N = 7630 N.

☐☐ For the previous problem, what must be the tension in the rope holding the balloon in place as in the figure?

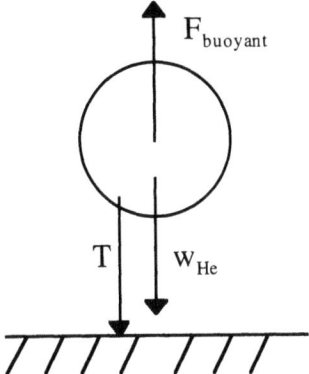

Ignoring the weight of the balloon material and the equipment,
$\Sigma F_{up} = \Sigma F_{down}$
$F_{buoy} = w_{He} + T \Rightarrow$ 8850 N - 1220N = T = 7630 N

☐☐ For the balloon of the previous problem, let the balloon material and equipment weigh 2500 N. What would be the mass of the load that the balloon could just lift?

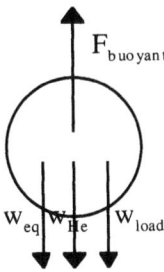

If the balloon were to just lift the load,
$\Sigma F_{up} = \Sigma F_{down}$
$F_{buoy} = w_{eq} + w_{He} + w_{load}$

8850 N = 2500 N + 1220 N + w_{load}

Thus, w_{load} = 5130 N.

☐☐ **An ideal fluid is flowing past a point in a pipeline at a rate of 6 gallons/min. What is the volume flow rate past any other point in the pipeline?**

The volume flow rate is constant for an ideal fluid. Thus at every other point in the pipeline the volume flow rate is 6 gallons per minute.

☐☐ **For the previous problem, how long would it take for a truck with a capacity of 1000 gallons to be filled from the pipeline?**

Since volume flow rate = volume/time

$\Rightarrow$ time = $\dfrac{volume}{volume\ flow\ rate}$

$\Rightarrow$ time = $\dfrac{1000\ gallons}{6\ gallons/\min}$ = 166 min = 2.78 hours

☐☐ **A small hole is drilled into a large tank (tank is open to atmosphere) 20 m below the surface of the liquid in the tank. With what speed will the liquid flow out of the tank?**

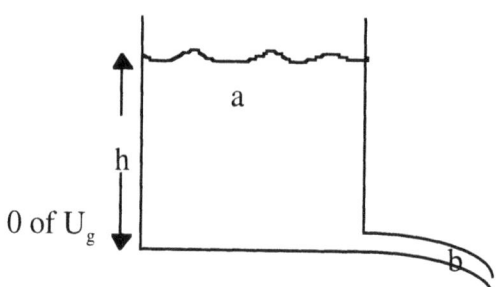

From Bernoulli's Theorem,

$P_a + (1/2)\rho v_a^2 + \rho g h_a = P_b + (1/2)\rho v_b^2 + \rho g h_b$

$P_{atm} + 0 + \rho g h = P_{atm} + (1/2) \rho v_b^2 + 0$

$\Rightarrow v_b^2 = 2gh$

$\Rightarrow v_b = \sqrt{2gh}$ (Torricelli's Theorem)

Thus, $v_{efflux} = \sqrt{2gh} = \sqrt{2(9.8 m/s^2)20m} = 19.8 m/s$

☐☐ **For the previous problem, if the area of the opening in the tank is 2cm², what is the rate at which the liquid comes out of the tank in (a) cubic meters per second, and (b) gallons/minute? (1 gallon = 3.79x10⁻³ m³)**

(a) volume flow rate = Av   (1cm² = 1x10⁻⁴ m²)
= (2x10⁻⁴ m²)(19.8 m/s) = 3.96x10⁻³ m³/s

(b) volume flow rate = $3.96 \times 10^{-3} \left(\dfrac{m^3}{s}\right)\left(\dfrac{1\ gallon}{3.79 \times 10^{-3} m^3}\right)\left(\dfrac{60s}{1\min}\right) = 62.7 \dfrac{gallons}{\min}$

☐☐ A 5 kg object is immersed in water and has an apparent weight of 32 N. Find (a) the volume of the water displaced and (b) the volume of the object.

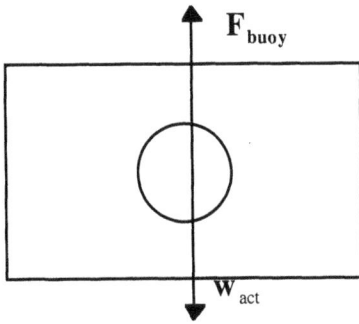

(a) From Archimede's Principle, an object immersed in a fluid is lifted up by a force equal to the weight of the fluid displaced. Thus,

$W_{apparent} = W_{actual} - F_{buoy}$

$W_{appz} = mg - \rho vg|_{H20}$

$32 N = (5kg)(9.8m/s^2) - (1000kg/m^3)(v)(9.8m/s^2)$

$\Rightarrow V_{H20 \; displaced} = 1.73 \times 10^{-3} m^3$

(b) $V_{H20 \; displaced} = V_{HBJ} = 1.73 \times 10^{-3} m$

☐☐ For the previous problem, find the mass density and the specific gravity of the object.

$\rho = \dfrac{m}{v} = \dfrac{5kg}{1.73 \times 10^{-3} m^3} = 2890 kg/m^3$

$S.G. = \dfrac{\rho_{obj}}{\rho_{H20}} = \dfrac{2880 kg/m^3}{1000 kg/m^3} = 2.89$

☐☐ A piece of wood is 0.75 m long, 0.25 m wide, and 0.05 m thick. If the density of wood is 600 kg/m³, what does the wood weigh?

$F = \rho vg = (600kg/m^3)(0.75 \; m)(0.25 \; m)(0.05 \; m)(9.8 m/s^2)$
$F = 55.1 \; N$

☐☐ The piston in a hydraulic car hoist is 0.15 m in radius. What fluid pressure is required to just start to lift a car of mass 1500 kg?

$P_{fluid} = \dfrac{F}{A} = \dfrac{(1500 kg)(9.8 m/s^2)}{\pi (0.15 m)^2}$

$P_{fluid} = 2.08 \times 10^5 \; N/m^2 = 2.06 \; atm$

☐☐ Water is flowing through a pipe at a rate of 2m³/min. Find the velocity of the fluid where the diameter of the pipe is (a) 7 cm and (b) 1aa5 cm.

(a) volume flow rate = Av

$$\frac{2m^3}{min}\left(\frac{1\,min}{60s}\right) = \pi(.035m)^2 v$$

v = 8.67 m/s where the pipe has a diameter of 7 cm.

(b) here $A_1 v_1 = A_2 v_2$

Thus, $\pi(.035m)^2 v_1 = \pi(.075m)^2 v_2$

v = 1.89 m/s where the pipe has a diameter of 15 cm.

☐☐ A jeweler orders a gold necklace that has a mass of 0.5 kg. When it arrives, the volume of the necklace is found to be 170 cm³. Is the necklace pure gold? (density of gold is 19,300 kg/m³)

$$\rho = \frac{m}{v} = \frac{0.5kg}{170 \times 10^{-6} m^3} = 2940 kg/m^3$$

Thus, the necklace is not made of gold.

☐☐ A small fraction of an iceberg protrudes above the surface of the ocean, while most of the iceberg is below the surface. Given that the density of ice is 920 kg/m³, while that of seawater is 1025 kg/m³, find the percentage of the iceberg's volume that lies above the surface.

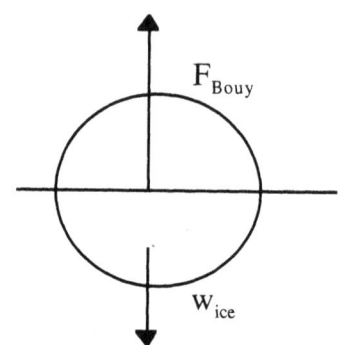

If iceberg is floating, then it's in a condition of equilibrium. Thus, $F_{bouy} = W_{ice}$, or

$$\rho V g_{seawater} = \rho g V_{ice} \quad \text{or} \quad \frac{V_{seawater}}{V_{ice}} = \frac{\rho_{ice}}{\rho_{seawater}} = \frac{920 kg/m^3}{1025 kg/m^3} = .898$$

Thus, volume of ice above ocean = 10.2 %.

☐☐ A fountain sends a stream of water 6 m into the air. Ignoring air resistance and friction, what must be the speed of the water leaving the fountain? Assume at that point the pressure is atmospheric.

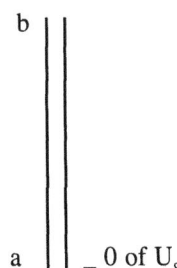

From Bernoulli's Theorem, $P_a + \frac{1}{2}\rho v_a^2 + \rho g h_a = P_b + \frac{1}{2}\rho v_b^2 + \rho g h_b$

$P_{atm} + \frac{1}{2}\rho v_a^2 + 0 = P_{atm} + 0 + \rho g h \Rightarrow v_a^2 = 2gh$

$v_a = \sqrt{2(9.8 m/s^2)(6m)}$

$v_a = 10.9$ m/s

☐☐ **An object has a mass m_1 in air and a mass m_2 in water. Show that the specific gravity of the object is given by S.G. = $\dfrac{1}{1 - \dfrac{m_2}{m_1}}$**

Let w_1 = weight in air and w_2 = weight in water. Then when object is in water,
1. $w_2 = w_1 - W_{water\ displaced} = m_2 g$ and when object is in air
2. $w_1 = m_1 g = \rho v g$.

Thus,
1. $m_2 g = \rho v g|_{obj} - \rho v g|_{water}$
2. $m_1 g = \rho v g|_{obj}$ or, $\dfrac{m_2}{m_1} = 1 - \dfrac{\rho_{H2O}}{\rho_{obj}} \Rightarrow \dfrac{\rho_{H2O}}{\rho_{obj}} = 1 - \dfrac{m_2}{m_1}$

or $\dfrac{\rho_{obj}}{\rho_{H2O}} = \dfrac{1}{1 - \dfrac{m_2}{m_1}}$ Hence, S.G. of object = $\dfrac{1}{1 - \dfrac{m_2}{m_1}}$

☐☐ **If an object has a mass of 25 g in air and a mass of 22 g in water, find its specific gravity and its density.**

From the previous problem,

S.G. = $\dfrac{1}{1 - \dfrac{m_2}{m_1}} = \dfrac{1}{1 - \dfrac{22g}{25g}}$

S.G. = $\dfrac{\rho_{obj}}{\rho_{H2O}} = 8.33$

$\Rightarrow \rho_{obj} = 8.33 \rho_{H2O} = 8330 kg/m^3$

☐☐ Air is flowing over an airplane's wings such that the speed is 45 m/s over the top surface and 28 m/s past the bottom surface. If the wings have an area of 12 m², and the plane has a mass of 750 kg, find (a) the lift force that acts on the aircraft and (b) the resultant force that acts on the aircraft. (Assume the density of air is 1.29 kg/m³.)

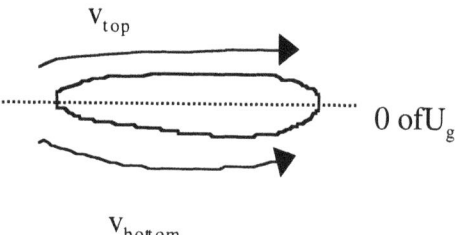

(a) Here $\Delta P = \dfrac{F}{A} \Rightarrow F = (\Delta P)A$

Using Bernoulli's theorem: $P_{top} + \dfrac{1}{2}\rho v_{top}^2 + \rho g h_{top} = P_{bottom} + \dfrac{1}{2}\rho v_{bottom}^2 + \rho g h_{bottom}$

Thus, $\dfrac{1}{2}\rho\left(v_{top}^2 - v_{bottom}^2\right) = P_{bottom} - P_{top} = \Delta P$

$\dfrac{1}{2}(1.29 kg/m^3)\left([45 m/s]^2 - [28 m/s]^2\right) = \Delta P$

$800 N/m^2 = \Delta P = 800 Pa$

Thus, F = (ΔP)A = (800 N/m²)(12 m²) = 9600 N

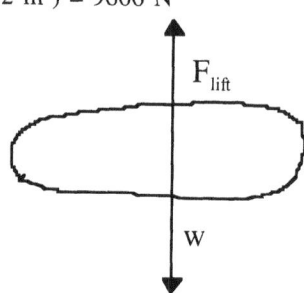

(b) $F_{resultant} = F_{lift} - w$ = 9600 N - (750 kg)(9.8 m/s²)

$F_{resultant}$ = 2250 N in the up sense

☐☐ An ore specimen weighs 16 N in air and 9.5 N in water. Find the volume and density of the specimen. (The density of water is 1000 kg/m³)

$W_{app} = W_{actual} - W_{water\ displaced}$

9.5 N = 16 N - $\rho v g|_{H20}$

$\dfrac{6.5 N}{(1000 kg/m^3)(9.8 m/s^2)} = v_{H20\ displaced} = 6.63 \times 10^{-4} m^3$

since $\rho = \dfrac{M}{v} \Rightarrow \rho = \dfrac{16 N / 9.8 m/s^2}{6.63 \times 10^{-4} m^3} = 2460 kg/m^3$

❏❏ Water is flowing in a circular pipe with a flow rate of 1 m³/s. What is the velocity of the water past a point where the radius of the pipe is 0.25 m?

Since the volume flow rate is constant, volume flow rate = Av

$$\frac{1 m^3/s}{\pi(.25m)^2} = v = 5.1 m/s$$

❏❏ For the flow tube of the figure, $A_1 = 4$ cm², $v_1 = 12$ m/s, and $A_2 = 1.5$ cm². Find (a) the volume flow rate and (b) the velocity of the fluid past point 2.

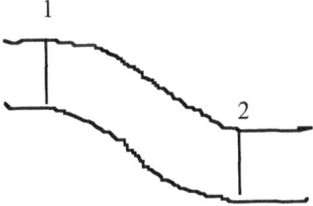

(a) Volume flow rate = $A_1 v_1 = (4 \times 10^{-4} m^2)(12 m/s)$.
Thus, the volume flow rate = $4.8 \times 10^{-3} m^3/s$.

(b) For an ideal fluid, the volume flow rate is constant. Thus, $A_1 v_1 = A_2 v_2$
$4.8 \times 10^{-3} m^3/s = (1.5 \times 10^{-4} m^2) v_2$
$v_2 = 32$ m/s

An important idea for fluids in motion is that where the _____ is greater, the _____ is less.

velocity, pressure.

❏❏ An aircraft has a mass of 10,400 kg and is in level flight. Find the pressure difference between the top and bottom surfaces of the wings of the airplane, if the wing area is 92 m².

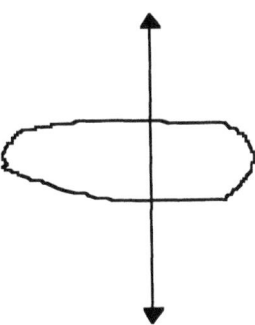

If the plane is in level flight, $F_{lift} = mg = (10,900 kg)(9.8 m/s^2)$
$F_{lift} = 106,800$ N

Since $\Delta P = \frac{F}{A} = \frac{106,800 N}{92 m^2} = 1160 Pa$

□□ Consider the figure. Find the average force that the water exerts on the dam, if the water is 30 m deep and the area of the face of the dam is 1200m².

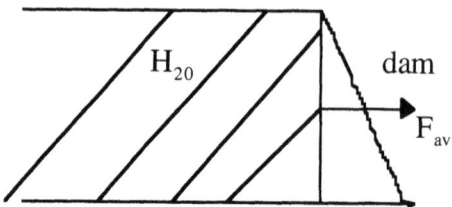

$$\Delta P = \frac{F}{A} \Rightarrow F_{av} = (\Delta P)_{av} A = \left(\frac{0 + P_{bottom}}{2}\right) A = \frac{\rho g h A}{2} = \frac{(1000 kg/m^3)(9.8 m/s^2)(30m)1200m^2}{2}$$

Thus, $F_{av} = 1.76 \times 10^8$ N

□□ A piece of gold-aluminum alloy weighs 50 N. When submerged in water, the apparent weight of the alloy is 35 N. What is the weight of the gold in the alloy? (The density of gold is 14,300 kg/m³ and of aluminum is 2700 kg/m³.

From Archimedes' Principle,
$F_{lift} = w_{actual} - w_{apparent} = 50$ N $- 35$ N

$F_{lift} = 15$ N $= \rho v g|_{H20}$

$$\frac{15N}{(1000 kg/m^3)(9.8 m/s^2)} = v_{H20} = v_{alloy} = 1.53 \times 10^{-3} m^3$$

Let x = volume of gold, then $1.53 \times 10^{-3} m^3$ - x = volume of aluminum.
Hence, $w_{gold} + w_{aluminum} = 50$ N
$\rho v g|_{gold} + \rho v g|_{aluminum} = 50$ N

$(19,300 kg/m^3)(x)(9.8 m/s^2) + (2700 kg/m^3)(1.53 \times 10^{-3} m^3 - x)(9.8 m/s^2) = 50$ N

OR

$(19,300 kg/m^3)x + 4.13$ kg $- (2700 kg/m^3)x = 5.1$ kg
$(16,600 kg/m^3)x = .97$ kg
$x = 5.84 \times 10^{-5} m^3$

Thus, $w_{gold} = \rho v g|_{gold} = (19,300 kg/m^3)(5.84 \times 10^{-5} m^3)(9.8$ m/s²$)$
$w_{gold} = 11.1$ N

As a check, $v_{aluminum} = 1.53 \times 10^{-3} m^3 - x$
$= 1.53 \times 10^{-3} m^3 - 5.84 \times 10^{-5} m^3$
$= 1.47 \times 10^{-3} m^3$

$w_{aluminum} = \rho v g|_{aluminum}$
$= (2700$ kg/m³$)(1.47 \times 10^{-3} m^3)(9.8$ m/s²$)$
$= 38.9$ N

Thus, 50 N $= w_{gold} + w_{aluminum}$
$= 11.1$ N $+ 38.9$ N $= 50$ N

☐☐ Consider the figure. If the area of piston 1 is 4 cm², what does the area of piston 2 have to be in order for the input force F_1 of 150 N to just lift a car of mass 3000 kg? (Assume both pistons are at the same level)

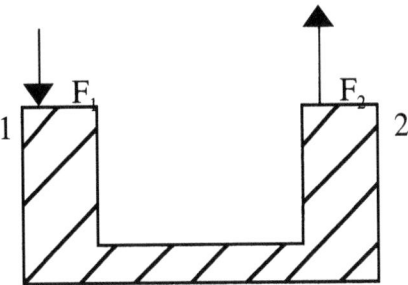

From Pascal's Principle, $\Delta p_1 = \Delta p_2$

$$\Rightarrow \frac{F_1}{A_1} = \frac{F_2}{A_2} \Rightarrow A_2 = \frac{A_1 F_2}{F_1}$$

$$= 4cm^2 \left(\frac{[3000kg][9.8m/s^2]}{150N} \right)$$

$$A_2 = 196 \text{ cm}^2$$

HEAT AND THERMODYNAMICS
TEMPERATURE SCALES AND EXPANSION

THE FUNDAMENTAL IDEAS ARE THE FOLLOWING:

- A thermometer is a device that is used to measure temperature.

- Thermometers can be liquid-in-glass, resistors, a stretched wire, or other physical quantity (called the thermometric property) that changes with temperature.

- Thermal equilibrium means that all parts of a system (that portion of the universe that is being examined) is at the same temperature.

- The Zeroth Law of Thermodynamics says that two systems in thermal equilibrium with a third are in thermal equilibrium with each other.

- Two systems in thermal equilibrium have the same temperature.

- Three common temperature scales are the following:
  a) Celsius Temperature Scale → freezing point of water at one atmosphere of pressure is 0°C and the boiling point of water at one atmosphere of pressure is 100°C.

  b) Fahrenheit Temperature Scale → freezing point of water at one atmosphere of pressure is 32°F and the boiling point of water at one atmosphere of pressure is 212°F.

  c) Absolute (Kelvin) Temperature Scale → here the degrees are the same size as on the Celsius Temperature Scale, but the zero is shifted so that 0K = -273.15 C°, i.e., T(K) = t°(c) + 273. Note that temperature on this scale has a unit of kelvins, e.g., 293 kelvins. Kelvin is capitalized only when referring the Kelvin Temperature Scale.

- A Δt of 100 Celsius degrees ≡ a Δt of 180 Fahrenheit degrees ⇒ Δt of 1° C ≡ Δt of 9/5° F.

- The easiest way to convert from °C to °F is to use the ratio $\dfrac{C}{100} = \dfrac{F-32}{180}$

- The change in length of a rod, wire, or cable because it experiences a change in temperature is given by $\Delta l = \alpha l_0 \Delta t$, where Δl is change in length, l_0 is original length, Δt is change in temperature, and α is the coefficient of linear expansion.

- The change in the area that comes about because a surface experiences a change in temperature is given by $\Delta A = 2\alpha A_0 \Delta t$, where ΔA = change in area, A_0 is original area, Δt is change in temperature, and 2α is the coefficient of area expansion.

- The change in volume that comes about because a volume (either liquid or solid) experiences a change in temperature is given by $\Delta V = \beta V_0 \Delta t$, where ΔV is change in volume, V_0 is original volume, Δt is change in temperature, and β is the coefficient of volumetric expansion. In the case of a solid, $\beta = 3\alpha$.

- Note that if an area has a hole in it, or a volume has a cavity, the expansion is the same as if the hole or cavity was filled with the material that surrounds it.

- Young's Modulus is given by $Y = \dfrac{\frac{F}{A}}{\frac{\Delta l}{l}}$

- Thus, $\dfrac{\Delta l}{l} = \dfrac{F}{AY}$. From linear expansion, $\dfrac{\Delta l}{l} = \alpha \Delta t$.

- Thus, $\alpha \Delta t = \dfrac{F}{AY} \Rightarrow \dfrac{F}{A} = Y\alpha \Delta t$ = Young's Modulus stress.

- If the rod, wire or cable is not free to move, if Δt is negative, the force is a positive tensile (stretching) force. If Δt is positive, the negative force is compressive. For these reasons, one writes that the stress is $\dfrac{F}{A} = -Y\alpha \Delta t$.

SOLVED PROBLEMS

□□ What is 27°C on the Fahrenheit and Kelvin temperature scales?

To convert to Fahrenheit,

$$\dfrac{F-32}{180} = \dfrac{C}{100} \Rightarrow \dfrac{F-32}{180} = \dfrac{27}{100} \Rightarrow F = \dfrac{(27)(180)}{100} + 32$$

F = 80.6°F

To convert to the Kelvin temperature scale: T(K) = t(°C) + 273
= 27 + 273
T(K) = 300 kelvins

□□ At what temperature, if ever, do the Celsius and Fahrenheit temperature scales read the same?

$\dfrac{C}{100} = \dfrac{F-32}{180}$ If F = C, then $\dfrac{F}{100} = \dfrac{F-32}{180}$

⇒ 180F = 100F - 3200
⇒ 80F = -3200
F = C = -40°

☐☐ **At what temperature, if ever, do the Fahrenheit and Kelvin temperature scales read the same?**

$T(K) = t°(C) + 273$ and $\dfrac{C}{100} = \dfrac{F - 32}{180}$

thus, $\dfrac{K - 273}{100} = \dfrac{F - 32}{180}$ and if $K = F$,

then $\dfrac{F - 273}{100} = \dfrac{F - 32}{180} \Rightarrow 180F - 273(180) = 100(F - 32)$

$80F = 45,940$
$F = K = 574.25$
thus, when both temperature scales read $574.25 \Rightarrow 574.25$ kelvins $= 574.25°$ F

☐☐ **If a 50 m long sidewalk (assumed to be one piece) expands by 0.015 m when the temperature changes from 2 to 38° C, find the coefficient of linear expansion for the concrete that was used to make the sidewalk.**

$\Delta l = \alpha l_0 \Delta t \Rightarrow \alpha = \dfrac{\Delta l}{l_0 \Delta t} = \dfrac{.015m}{50m(36°C)} \Rightarrow \alpha = 8.33 \times 10^{-6} / °C$

☐☐ **By how much would the sidewalk of the previous problem decrease in length if the temperature changed from 105° F to -20° F?**

(Method 1: change Δt from ° F to ° C):
$\Delta t = (-20° F - 105° F) = -125°$ F
Δt of $1°$ C $= \Delta t$ of $9/5°$ F
Thus, Δt of $(-125° F)\left(\dfrac{1°C}{9/5°F}\right) = -69.4°C$
Thus, $\Delta l = \alpha l_0 \Delta t = (8.33 \times 10^{-6}/°C)(50m)(-69.4° C)$
$\Delta l = -2.89 \times 10^{-2}$ m

(Method 2: convert each temperature from Fahrenheit to Celsius degrees):
$\dfrac{C}{100} = \dfrac{F - 32}{180} \Rightarrow \quad C = 5/9(F-32)$
$\quad C = 5/9(105-32) = 40.5°$ C
$\quad C = 5/9(-20-32) = -28.9°$
Thus, $\Delta t = -69.4°$ C $\Rightarrow \Delta l = -2.89 \times 10^{-2}$ m, as before.

(Method 3: convert the coefficient of linear expansion from a #/° C to a #/° F):
$\alpha = 8.33 \times 10^{-6} / °C \left(\dfrac{1°C}{9/5°F}\right) = 4.63 \times 10^{-6} / °F$

Thus, $\Delta l = \alpha l_0 \Delta t$
$= (4.63 \times 10^{-6}/°F)(50m)(-125° F)$
$= -2.89 \times 10^{-2}$ m

☐☐ **The normal boiling point of liquid nitrogen is 77 kelvins. What is this temperature on the Celsius temperature scale? On the Fahrenheit temperature scale?**

$T(K) = t (° C) + 273 \Rightarrow t (° C) = T(K) - 273 = 77 - 273 = -196 °$ C

$$\frac{C}{100} = \frac{F-32}{180} \Rightarrow \frac{180}{100}C + 32 = F$$

$$\frac{180}{100}(-196) + 32 = F = -321°F$$

◻◻ A steel bridge is built on a hot summer day when the temperature is 35 ° C. What is the new length of the bridge on a cold winter day when the temperature is -15 ° C if the original length is 90 m? (Coefficient of linear expansion for steel is 1.2×10^{-5}/ ° C)

$\Delta l = \alpha l_0 \Delta t$
   $= (1.2 \times 10^{-5}$ / ° C$)(90$ m$)(-15$ ° C $- 35$ ° C$)$
$\Delta l = -.054$ m

Thus, the new length would be 89.95 m

◻◻ For the previous problem, what would be the stress that acts on the girders of the bridge when the temperature is -15 ° C (Young's Modulus for steel is 2×10^{11} N/m²)

$F/A = -y\alpha \Delta t = -2 \times 10^{11}$ N/m² $(1.2 \times 10^{-5}$/ ° C$)(-50$ ° C$)$
            $= 1.2 \times 10^8$ N/m²

◻◻ A glass beaker has a capacity of 1000 cm³ at 5 ° C. What is its volume at 100 ° C if the coefficient of linear expansion of glass is 1.3×10^{-5}/ ° C?

$\Delta V = \beta V_0 \Delta t$. For a solid $\beta = 3\alpha$. Thus, $\Delta V = 3(1.3 \times 10^{-5}$/ ° C$)(1000$ cm³$)(95$ ° C$)$
$\Delta V = 3.71$ cm³
The new volume at 95 ° C is 1003.7 cm³.

◻◻ A steel gas tank holds 25 gallons and is filled when the temperature is 60 ° F. Later, the temperature rises to 105 ° F. How much gasoline, if any, runs out?
($\alpha_{steel} = 1.2 \times 10^{-5}$/ ° C, $\beta_{gasoline} = 1.2 \times 10^{-3}$/ ° C)

$\Delta t = 105°F - 60°F = 45°F \left(\frac{(5/9)°C}{1°F} \right) = 25°C$

$\Delta V_{gasoline} = \beta V_0 \Delta t = (1.2 \times 10^{-3}$ / °C$)(25 gal)(25°C)$
$\Delta V_{gasoline} = .75 gallons$

$\Delta V_{steel} = \beta V_0 \Delta t = (3\alpha) V_0 \Delta t = (3.6 \times 10^{-5}$ / °C$)(25 gal)(25°C)$
$\Delta V_{steel} = .023 gallons$

Amount out = $\Delta V_{gasoline} - \Delta V_{steel} = (.75 - .023)$ gallons

Amount out = .727 gallons

◻◻ Steel railroad rails 20 m long are laid when the temperature is -8° C. (a) How much space must be left between rails if the rails are just to touch on a summer day when the temperature is 35 ° C. and (b) if the rails touched (were in contact) when laid, what would be the stress in them when the temperature is 35 ° C? ($\alpha_{steel} = 1.2 \times 10^{-5}$ / °C, Young's Modulus for steel = 2×10^{11} Pa)

(a) $\Delta l = \alpha l_0 \Delta t$
$= (1.2 \times 10^{-5}/°C)(20\text{ m})(43°)$
$\Delta l = .0103\text{ m} = 10.3\text{ mm}$

(b)
$\frac{F}{A} = -Y\alpha\Delta t$
$= -(2 \times 10^{11} N/m^2)(1.2 \times 10^{-5}/°C)(43°C)$
$= -1.03 \times 10^8 N/m^2 \left(-1.03 \times 10^8 N/m^2\right)$

The – means the force is compressive.

☐☐ An aluminum cube 0.11 m on a side is heated from 5 to 55 °C. Find (a) its change in volume, and (b) its change in density. ($\beta_{aluminum}$ = 7.2 x 10^{-5}/ °C, density of aluminum = 2700 kg/m³)

(a) $\Delta V = \beta V_0 \Delta t$
$= (7.2 \times 10^{-5}/°C)(.11m)^3(50°C)$
$= 4.79 \times 10^{-6} m^3$

(b) $\rho = \frac{M}{V} \Rightarrow m = \rho V = (2700 kg/m^3)(.11m)^3$

mass of cube = 3.59 kg
$\Delta l|_{edge} = \alpha l_0 \Delta t = (2.4 \times 10^{-5}/°C)(.11m)(50°C)$
$\Delta l|_{edge} = 1.32 \times 10^{-4}$ m

$l_{new} = (.11 + 1.32 \times 10^{-4})$m
$\Delta\rho = \rho_{new} - \rho_{old} = \frac{m_{new}}{V_{new}} - \frac{m_{old}}{V_{old}}$

$= 3.59 kg \left(\frac{1}{(.110132m)^3} - \frac{1}{(.11m)^3}\right)$

$\Delta\rho = -9.69$ kg/m³. [kg/m³]
This implies the density decreases.
[Note: this problem is easier using calculus which finds that $\Delta\rho = (-\rho\Delta V)/V$

So $\Delta\rho = \frac{-2700 kg/m^3 (4.79 \times 10^{-6} m^3)}{(.11m)^3}$

$\Delta\rho = -9.71$ kg/m³

☐☐ A wire that is 3.5 m long at 15° C increases in length by 2.1 cm when heated to 480°C. Find the average coefficient of linear expansion.

$\Delta l = \alpha l_0 \Delta t \Rightarrow .021 m = \alpha(3.5m)(480-15)°C$

$\alpha = 1.29 \times 10^{-5}/°C$

⬜⬜ **A pendulum clock keeps perfect time when the temperature is 23° C. How much time (in seconds) would the clock gain or lose in a day if the temperature fell to 10° C? Assume motion is SHM, and that the pendulum is made of brass which has a coefficient of linear expansion of $2 \times 10^{-5}/°C$.**

For SHM, $T = 2\pi \sqrt{\dfrac{l}{g}}$

Thus, T_2 (period when the temperature is 10° C) and T_1 (period when the temperature is 23° C) can be written as

$$T_2 = 2\pi \sqrt{\dfrac{l_2}{g}} \quad \text{and} \quad T_1 = 2\pi \sqrt{\dfrac{l_1}{g}}$$

OR $\dfrac{T_2^2}{T_1^2} = \dfrac{l_2}{l_1} = \dfrac{l_1 + \alpha l_1 \Delta t}{l_1} = 1 + \alpha \Delta t$

so that $T_2^2 = T_1^2 + T_1^2 \alpha \Delta T$
or $T_2^2 - T_1^2 = T_1^2 \alpha \Delta t$
$(T_2 + T_1)(T_2 - T_1) = T_1^2 \alpha \Delta t$

$T_2 - T_1 = \Delta T = \dfrac{T_1^2 \alpha \Delta t}{T_2 + T_1} = \dfrac{T_1^2 \alpha \Delta t}{2T_1}$ since $T_2 \approx T_1$

Finally, $\Delta T = \dfrac{T_1 \alpha \Delta t}{2} = \dfrac{(86,400s)(2 \times 10^{-5}/°C)(-13°C)}{2}$

$\Delta T = -11.2s$ for one day.
Since the pendulum is shorter at the lower temperature, the clock gains 11.2s / day.

⬜⬜ **For the previous problem, how closely must the temperature be controlled so that the clock is not to gain or lose more than 2 s per day?**

$\Delta T = \dfrac{T_1 \alpha \Delta t}{2}$

$\dfrac{2\Delta T}{T_1 \alpha} = \Delta t = \dfrac{2(\pm 2s)}{(86,400s)(2 \times 10^{-5}/°C)}$

$\Delta t = \pm 2.31°C$

⬜⬜ **A glass beaker has a volume of 500 cm³ at 0° C and is completely filled with mercury ($\beta = 18 \times 10^{-5}/°C$). When the glass beaker and mercury are heated to 100 ° C, 6.7 cm³ of Hg overflow. Find the coefficient of volume expansion of the glass of which the beaker is made.**

Amount out $= \Delta V_{Hg} - \Delta V_{glass} = \beta V_0 \Delta t_{Hg} - \beta V_0 \Delta t_{glass}$

$6.7 cm^3 = (18 \times 10^{-5}/°C)(500 cm^3)100°C - \beta(500 cm^3)100°C$
$6.7 cm^3 = 9 cm^3 - \beta(500 cm^3)(100°C)$

$\beta = \dfrac{2.3 cm^3}{(500 cm^3)(100°C)} = 4.6 \times 10^{-5}/°C$

☐☐ At 20° C, a glass flask (200 cm³) is filled completely to a mark on the long stem of the flask. (β of liquid is 40 x 10⁻⁵/ ° C, β of glass is 2.5 x 10⁻⁵/ ° C). If the cross-sectional area of the stem is 4 mm², how far up the stem will the liquid rise (or fall) when the temperature is raised to 38 ° C? Ignore the expansion of the stem.

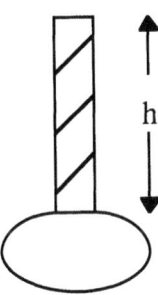

Volume (up or down) in stem = Ah

Thus, $Ah = \Delta V|_{liquid} - \Delta V|_{glass}$

$Ah = \beta V_0 \Delta t|_{liquid} - \beta V_0 \Delta t|_{glass}$

$Ah = (40 \times 10^{-5}/° C)(100 cm^3)18° C - 2.5 \times 10^{-5}/° C (100 cm^3)18° C$

$Ah = .72 cm^3 - .045 cm^3 = .675 cm^3$

$h = \dfrac{.675 cm^3}{4 mm^2 \left(\dfrac{1 cm^2}{1 \times 10^{-2} mm^2} \right)} = 16.9 cm$

The liquid rises up the stem.

HEAT

THE FUNDAMENTAL IDEAS ARE THE FOLLOWING:

- Heat (symbol Q) is energy in transit because of temperature differences.

- The unit of heat energy is joules, but the following units have also been used:
  a. Calorie - the amount of heat needed to raise the temperature of one gram of water from 14.5 to 15.5 °C.
  b. Kilocalorie - the amount of heat needed to raise one kilogram of water from 14.5 to 15.5 °C.
  c. Food Calorie - the same as a kilocalorie, denoted by Calorie spelled with a capital C.
  d. British thermal unit (BTU) - the amount of heat needed to raise the temperature of one pound of water from 63 to 64 °F.
  e. Therm - 100,000 BTU's.
  f. Ton - the amount of heat needed to change one ton of water at 32 °F to one ton of ice at 32 °F.

- Heat capacity, C, is defined to be $\frac{\Delta Q}{\Delta t}$.

- Specific heat, $c = C/M \Rightarrow \Delta Q = mc\Delta t$.

- The usual sign convention is that when heat enters an object it's positive, when heat leaves an object it's negative.

- For a gas $m = nM$, where n = number of moles and M = molecular weight (mass)

- Thus, $Q = nMc\Delta t$, or $Q = nc_n \Delta t$ where c_n = molar heat capacity or $c_n = Mc$

- When matter changes phase, i.e., undergoes a change from a solid to a liquid, or from a liquid to a gas, or vice-versa, heat is either added or extracted.

- If there is no phase change, the amount of heat needed to change the temperature of an object is given by $Q = mc\Delta t$.

- The following are useful definitions:
  a. Melting point - that temperature at which the solid changes to the liquid state.
  b. Freezing point - that temperature at which the liquid changes to the solid state.
  c. Latent heat of fusion (L_f) - the amount of heat needed to change a unit mass of a solid to the liquid at the same temperature and pressure.
  d. Boiling point - 1.- that temperature at which vapor bubbles form in the liquid and rise to the surface.
  2.- that temperature at which the saturated vapor pressure equals the applied pressure.
  e. Latent heat of vaporization (L_V) - the amount of heat needed to change the liquid to the vapor state at the same temperature.
  f. Evaporation - escape of molecules from the surface of a liquid.

- Note that the heat necessary to cause a phase change is given by $Q = mL$ where m is mass and L is either the heat of fusion or the heat of vaporization.

- Note that a system is that portion of the universe under examination, and that the surround is that part of the universe that surrounds the system. Thus, the universe equals the system plus the surround.

- If the system is insulated so that no heat can enter or leave the system, then if a hot object (or objects) is placed in the system, the hot object (or objects) lose heat and the cold object (or objects) gain heat until all parts of the system are at the same temperature. Thus, the heat lost by those objects losing heat equals the heat gained by those objects gaining heat. This is called calorimetry, i.e. $Q_{lost} = Q_{gained}$.

- Under ordinary circumstances, if heat is added to a solid it changes to a liquid, if more heat is added, the liquid changes to a gas. Under some conditions a solid can change directly to the gaseous phase. This transition from the solid to the vapor state is called sublimation.

- Heat of combustion is the energy per unit volume or per unit mass released by oxidation (burning).

SOLVED PROBLEMS

❑❑ **Given that the specific heat of aluminum is 900 J/kg°C, how much heat must be added to a 12 kg block of aluminum to raise its temperature from 15 to 60 °C?**

If there is no phase change, $Q = mc\Delta t$
$Q = 12$ kg $(900$ J/kg°C$) (45$ °C$) = 4.86 \times 10^5$ J

❑❑ **From the definition of a British thermal unit (BTU), find how many Joules are in one BTU. (1 pound = .454 kg, specific heat of water = 4186 J/kg°C)**

$Q = mc\Delta t$
one BTU is the amount of heat that will raise one pound of water one degree Fahrenheit.
Thus, 1 BTU = .454 kg (4186 J/kg°C)(5/9 °C) ≅ 1055J

❑❑ **800 Joules of heat are added to 160 g of a liquid, and its temperature rises by 6.5 °C. Find the specific heat of the liquid.**

$Q = mc\Delta t \Rightarrow 800$ J $= (.16$ kg$)(c)(6.5$ °C$)$
$c = 769$ J/kg °C

❑❑ **How much energy would it take to change 5 kg of ice at -30 °C to steam at 100 °C? ($c_{H20} = 4186$ J/kg °C, $c_{ice} = 2100$ J/kg°C, $L_f = 335 \times 10^3$ J/kg, $L_v = 2260 \times 10^3$ J/kg)**

$Q = mc\Delta t|_{ice} + mL_f + mc\Delta t|_{H20} + mL_v =$
$(5kg)(2111 J/kg°C)(30°C) + 5kg(335x10^3 J/kg) + 5kg(4186 J/kg°C)(100°C) + 5kg(2260x10^3 J/kg) Q = 1.54 \times 10^7$ J

❏❏ An aluminum pan (mass = 200 g, c = 900 J/kg° C) holds 300 g of water (c_{water} = 4186 J/kg° C) at 5° C. What power electric heater is needed to bring the pan and the water to a boil (100° C) in 3 minutes? (Assume no heat lost to surround.)

$$P_{av} = \frac{\Delta E}{time} = \frac{Q}{time} = \frac{mc\Delta t|_{pan} + mc\Delta t|_{water}}{time}$$

$$P_{av} = \frac{(.2kg)(900J/kg°C)95°C + .3kg(4186J/kg°C)(95°C)}{180s}$$

P_{av} = 758 J/s = 758 watts

❏❏ A 1500 kg car is braked to a halt from 80 mi/hr. 80% of the car's initial kinetic energy appears as heat in the steel rotors of the brake system (rotors have a total mass of 12 kg), What is the increase in temperature of the rotors? (c_{steel} = 450 J/kg° C, 1 mile = 1609m)

$.8(1/2mv^2)_{car} = mc\Delta t|_{rotors}$
$.8(1/2)(1500kg)(35.8m/s)^2$ = 12 kg(450 J/kg° C)Δt or Δt = 142 ° C

❏❏ Given that the specific heat of air is 1020 J/kg° C, and that one L of air has a mass of 1.29 x 10^{-3} kg, find (a) the amount of heat needed to raise the temperature of the 0.5 L of air inhaled with each breath from - 35° C to body temperature of 37 ° C, and (b) how much heat is lost in one hour if the respiration rate is 15 breaths per minute.

(a) Q = mcΔt
 = (.5)(1.29 x 10^{-3} kg)(1020 J/kg° C)(72° C)
 Q = 47.4 J/breath

(b) Here, each breath requires 47.4 J, and if there are 15 breaths/min (60 min) ⇒ that there are 900 breaths in one hour.
Thus, Q = 47.4J/breath (900 breaths) and Q = 4.26 x 10^4 J

❏❏ A 50 g specimen at a temperature of 98° C is dropped into 100 g copper calorimeter that contains 200 g of water at 20° C. The final temperature of the calorimeter is 23.6° C. Find the specific heat of the specimen. (Specific heat of copper is 390 J/kg° C, and the specific heat of water is 4186 J/kg°C)

The fundamental idea of calorimetry is that heat lost equals heat gained.
Thus, $mc\Delta t|_{speciman} = mc\Delta t|_{calorimeter} + mc\Delta t|_{water}$

$(.05kg)C(74.4°C) = .1kg(390J/kg°C)(3.6°C) + (.2kg)(4.86J/kg°C)(3.6°C)$

c = 848J/kg° C

Note that in using this method the changes in temperatures are made positive.

An alternative approach is to recognize that if no heat enters or leaves the calorimeter,

$0 = Q_{gained} + Q_{Lost}$

$0 = mc\Delta t|_{calorimeter} + mc\Delta t|_{water} + mc\Delta t|_{speciman}$

so that $-mc\Delta t|_{speciman} = mc\Delta t|_{calorimeter} + mc\Delta t|_{water}$

$-(.05kg)(c)(23.6°C - 98°C) = .1kg(390J/kg°C(23.6 - 20)°C + .2kg(4186J/kg°C(23.6 - 20)°C$

⇒ $(.05kg)(c)74.4°C = (.1)(390)3.6J + .2(4186)(3.6)J$

or $c_{specimen}$ = 848J/kg° C as before.

☐☐ A copper calorimeter (specific heat 390 J/kg° C) that has a mass of 350 g contains 50 g of ice, and the calorimeter and the ice are at 0° C. 20 g of steam at 100° C enters the calorimeter. What is the final temperature of the calorimeter and its contents? (Specific heat of water is 4186 J/kg° C, L_f for water is 335 x 10^3 J/kg, L_v for water is 2260 x 10^3 J/kg)

Here the steam condenses and the ice melts. Thus,

0 = heat gained + heat lost

$0 = mc\Delta t|_{calorimeter} + mL|_{ice} + mc\Delta t|_{water\ that\ was\ ice} - mL_v|_{steam} + mc\Delta t|_{water\ that\ was\ steam}$

$0 = (.35kg)(390J/kg°C)(t_E - 0)°C + (.05kg)335x10^3 J/kg + .05kg(4186J/kg°C)(t_E - 0)°C$

- .02kg(2260 x 10^3J/kg) + .02kg(4186J/kg°C)(t_E-100)°C

t_E = 85.7° C

☐☐ A student uses a 200 watt immersion heater to heat 0.3 kg of water from 23 to 100 °C. find (a) how much heat must be added to the water, and (b) if all of the heat from the heater is absorbed by the water how long does it take? (Specific heat of water is 4186J/kg° C)

(a) Q = mcΔt
     = (.3kg)(4186J/kg° C)(77° C)
     Q = 9.67 x 10^4 J

(b) Power = $\frac{\Delta E}{time}$ ⇒ time = $\frac{\Delta E}{Power}$ = $\frac{9.67x10^4 J}{200J/s}$

time = 484s or 8.05 minutes

☐☐ Given that 1 pound = .454 kg, how many Joules are in a "ton" of heat? (L_f for water is 335 x 10^3J/kg)

A "ton": of heat is the amount of heat needed to change one ton (2000 lbs) of water at 32°F to ice at ice at 32°"F in 24 hours.

Thus, Q = mL

$= 2000lb \left(\frac{.454kg}{1lb} \right)\left(335x10^3 J/kg \right)$

Q = 3.04 x 10^8J

☐☐ For the previous problem, what power in watts is required?

$P_{av} = \frac{\Delta E}{time} = \frac{3.04x10^8 J}{86400s} = 3520 watts$

☐☐ A large block of ice is at 0° C. A 2 kg chunk of iron at 200° C is placed on the block (specific heat of iron is 450J/kg°C). How much ice is melted? (The heat of fusion of water is 335 J/g).

The heat from the iron melts some ice. Thus, Q = mcΔt$|_{iron}$ = 2 kg(450 J/kg°C)(200° C)

Q = 1.8 x 10^5J of heat is given off from the iron. Since for the phase change of ice to water at the same temperature, Q = mL$_f$ ⇒ $\frac{1.8x10^5 J}{335J/g}$ = 537g

Thus, .537 kg of ice is melted.

❏❏ A survival school trainee is climbing Punishment Mountain. If the trainee and his pack have a mass of 140 kg, how many grubs (2 food calories each) would the trainee have to eat to provide the energy for the climb? Punishment Mountain is 150 m high, and only 20% of the energy obtained from the grubs is available to provide the energy for the trainee to climb Punishment Mountain. (1 food calorie is 4186 J).

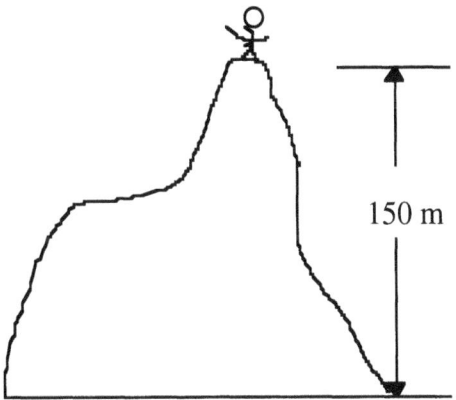

The idea here is ΔE = work done. The work done here is mgh,
or $(140 \text{ kg})(9.8 \text{ m/s}^2)(150 \text{ m})$ which equals 2.06×10^5 J. Since each grub has 2 food calories, or 8370 Joules, .2 (n) 8370 J) = 2.06×210^5 J where n equals the number of grubs. Thus, n ≅ 123 grubs.

❏❏ A 2 kg copper block is given an initial speed of 4 m/s on a surface where friction is present and comes to rest. The block absorbs 75% of its initial kinetic energy in the form of heat. Find the increase of temperature of the block. (The specific heat of copper is 390 J/kg° C)

$k_0 = (1/2)mv^2 = (1/2)(2 \text{ kg})(4 \text{ m/s})^2 = 16$ J
Since 75% of this energy is absorbed as heat, $Q = mc\Delta t \Rightarrow (0.75)(16 \text{ J}) = 2 \text{ kg}(390 \text{ J/kg° C})\Delta t$. Thus, the increase of temperature of the copper block is 0.15° C.

❏❏ A 1600 kg car has steel brake rotors of mass 50 kg. If all of the kinetic energy of the car is transferred to the brake rotors in the form of heat as the car is braked to a halt, what would be the final temperature of the brake rotors if the car is brought to a halt from 60 m/h (26.8 m/s)? Ignore the transfer of heat from the rotors. (The specific heat of steel is 450 J/kg° C, and the initial temperature of the rotors is 20° C)

The initial kinetic energy of the car is transformed to heat and thus raises the temperature of the brake rotors. Thus $(1/2)mv^2|_{car} = mc\Delta t|_{rotor}$ or $(1/2)(1600 \text{ kg})(26.8 \text{ m/s})^2 = 50 \text{ kg}(450 \text{ J/kg° C})\Delta t$., so that Δt = 25.5° C, and the final temperature of the brake rotors is 45.5° C.

❏❏ For the previous problem, if the melting point of steel is 1500° C, how many repeated stops would it take for the brake rotors to reach the melting point? Ignore the transfer of heat from the rotors.

Since one stop corresponds to a Δt of 25.5° C, $n(\Delta t) = 1500 - 20°$ C; under the conditions set out 58 stops could be made before the melting point of the brake rotors was reached.

HEAT TRANSFER

THE FUNDAMENTAL IDEAS ARE THE FOLLOWING:

- There are three methods of heat transfer:
  a. Conduction - the transfer of heat by means of molecular interaction.
                  - a material has to be present for conduction to take place.

  b. Convection - the transfer of heat by means of the movement of the heated material.
                  - a material has to be present for convection to take place.

  c. Radiation - the transfer of heat by means of electro-magnetic waves.
                  - no material has to be present for radiative transfer to occur.

- The model for conduction says that the heat conducted through a slab of material is given by $Q = \frac{kA(T_h - T_c)time}{L}$, where Q is heat, k is the coefficient of thermal conductivitiy, A is the surface area, $T_h - T_c$ is the temperature difference between the hot and cold sides, and L is the thickness of the slab of material.

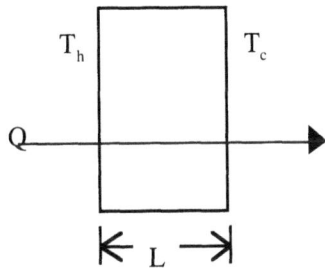

It may be convenient to write the conduction of heat as heat current H, where
$$H = \frac{Q}{time} = \frac{kA(T_h - T_c)}{L}$$

If there are several slabs of material, it can be shown that $H = \dfrac{A(T_h - T_c)}{\dfrac{L_1}{k_1} + \dfrac{L_2}{k_2} + \dfrac{L_3}{k_3} + ...}$

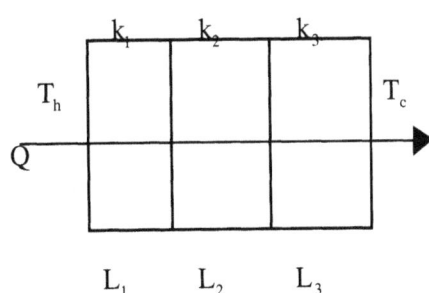

- The units of the coefficient of thermal conductivity are $\frac{J}{s\,m\,°C}$ or $\frac{watt}{m\,°C}$

- Note that the bigger the coefficient of thermal conductivity, the more heat transferred by conduction.

- The model that is used for the convection of heat says that the heat current, or convection current, is given by $H = hA(T_h - T_c)$, where h is called the convection coefficient and the other symbols have the usual meaning. If the convection current is caused by temperature differences, it is called natural convection; if the convection current is caused by a fan or a pump, it is called forced convection.

- Radiation is the transfer of energy by means of electro-magnetic waves and is usually measured in terms of energy per unit time per unit area. Radiant emittance (R) is the energy per unit time per unit area emitted by an object, while irradiance (H) is the energy per unit time per unit area that is incident upon an object. A black body is an object that absorbs all radiation incidents upon it for an object to remain at the same temperature, $R_B = H$. The Stefan-Boltzmann Law says that $R_B = \sigma T^4$, where σ is the Stefan-Boltzmann constant and equals 56.7×10^{-9} watts/m²k⁴, and T is the Kelvin (Absolute) temperature. A real object does not absorb all radiation incident upon it, and thus one defines emissivity $e = \frac{R}{R_B}$. Since an object both absorbs and radiates energy the heat transferred by radiation is given by $\frac{Q}{time} = H = R - H = A\varepsilon\sigma T^4 - A\varepsilon\sigma T_0^4$, where T = the temperature of the object and T_0 = the temperature of the surround. Note that a good absorber is a good radiator, thus, a dark colored object not only absorbs more energy, it also emits or radiates more energy.

SOLVED PROBLEMS

❏❏ What is the heat current through a glass window that has dimensions of 2 m by 1.6 m if the temperature difference between the hot side and the cold side is 15 °C and the glass is 3 mm thick? (The coefficient of thermal conductivity of glass is 0.8 watt/ m-°C)

$$H = \frac{kA(T_h - T_c)}{L} = \frac{\left(\frac{.8\,watt}{m\,°C}\right)(2m)(1.6m)(15°C)}{3 \times 10^{-3}m}$$

H = 12,8000 watts = 12,8000 J/s

❏❏ An oven has a surface area of 1.8 m², and the interior temperature is kept at 200 °C, while the outside temperature is 23 °C. The oven is lined with insulation that is 6 cm thick and whose thermal conductivity is 0.08 watt/m°C. What is the rate at which heat is flowing from the oven?

$$H = \frac{kA(T_h - T_c)}{L} = \frac{\frac{.08\,watt}{m\,°C}(1.8m^2)(177°C)}{.06m} = 425\,watts$$

Thus, heat is leaving the oven at the rate of 425 watts or 425 J/s.

❏❏ For the previous problem, at what rate must the oven element provide heat to maintain the oven at 200 °C?

For the temperature to remain constant, the oven element must provide heat at the same rate as the rate at which heat leaves the oven, i.e., 425 watts or 425 J/s.

❏❏ An aluminum pan that is 2 mm thick (area of bottom is 360 cm^2, coefficient of thermal conductivity 240 watt/m°C) is on a burner filled with boiling water at a temperature of 100 °C. If 110 g of water boils away in one minute, find the temperature of the bottom surface of the pan that is in contact with the burner.

$L_V = 2260 \times 10^3$ J/kg, thus, $Q = mL_V = (.11 \text{ kg})(2260 \times 10^3 \text{J/kg})$
$Q = 2.49 \times 10^5$ J in 60 s

$$Q = \frac{kA(T_h - T_c)\text{time}}{L} \Rightarrow \frac{QL}{(\text{time})kA} = T_h - T_c$$

so that $\dfrac{(2.49 \times 10^5 J)(2 \times 10^{-3} m)}{(60s)(240 J/sm°C)(360 \times 10^{-4} m^2)} = T_h - T_c = .96°C$

Thus, $T_h \approx 101$ °C.

❏❏ A black body is at 23 °C. What is its radiant emittance?

From the Stefan-Boltzmann Law, $R_B = \sigma T^4$
$\dfrac{56.7 \times 10^{-9} \text{watt}}{m^2 K^4} (300k)^4$
$R_B = 459$ watt/m^2

❏❏ The solar constant is the irradiance at the upper surface of the earth's atmosphere, and has been measured to be about 1400 watt/m^2. Assuming that the earth-sun distance is 1.4×10^{11} m, and that the sun is a black body, find the temperature of the sun. (the radius of the sun is 7×10^8 m).

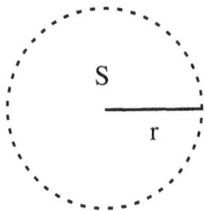

For a black body $H_B = R_B$. If the solar constant is 1400 watt/m^2, then the total energy per unit time emitted by the sun is 1400 watts/m^2 $(4\pi r^2)$ or 3.45×10^{26} watts.

$R_B = \dfrac{\text{energy}}{(\text{time})(\text{area})} \Rightarrow \dfrac{\text{energy}}{\text{time}} = (R_B)(\text{area of sun}) \Rightarrow 3.45 \times 10^{26}$ watts $= \sigma T^4 4\pi (r_s)^2$

Thus, 3.45×10^{26} watts $= (56.7 \times 10^{-9}$ watts/m^{2}K$^4)(T^4)(4\pi)(7 \times 10^8 m)^2$ or
$T^4 = 9.87 \times 10^{14}$ K^4
$T \cong 5600$k

☐☐ **A parked car on a hot summer day absorbs energy at a rate of 600 watt/m². After a while the car emits energy at the same rate. If the emissivity of the car is 0.6, what is the Celsius temperature of the car?**

$R = H \Rightarrow R = \varepsilon \sigma T^4$
600 watts/m² = $(0.6)(56.7 \times 10^{-4}$ watt/m²K⁴$)T^4$
$T = 364K^4$

$T(k) = t \, (°C) + 273$. Thus, the car's temperature is 91 °C.

☐☐ **Why should radiators on a car be painted black?**

Since a good absorber is a good radiator (emitter), the radiator will radiate more heat if it is black.

☐☐ **A brick outside wall of area 10 m² (10 cm thick) is next to sheet rock (1.5 cm thick). If the inside temperature is 27 °C and the outside temperature is 0 °C, (a) what is the heat current through the combination, and (b) what is the temperature of the interface (where the brick and sheetrock meet)? (The coefficient of thermal conductivity of brick and sheetrock are 0.6 J/(s-m°C) and 0.3 J/(s-m°C), respectively.**

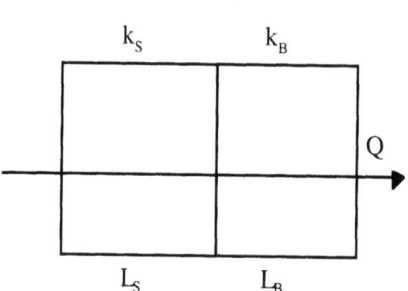

- Assuming steady state condition,

$$H = \frac{A(T_h - T_c)}{\frac{L_1}{k_1} + \frac{L_2}{k_2}} = \frac{10m^2(27°C - 0°C)}{\frac{.015m}{0.3 watt/m°C} + \frac{.1m}{0.6 watt/m°C}}$$

H = 1250 watts

- To find the temperature at the sheetrock-brick interface:

$H_S = H_B$ under steady state condition,

so $H_S = \frac{kA(T_h - T_I)}{L_S}$

$1250 \, J/s = \frac{(0.3J/s - m°C)(10m^2)(27° - T_I)}{.015M}$

Thus, 6.3 °C = 27° - T_I or T_I = 20.7 °C

☐☐ An object emits power at a rate of 40 watts. If it were a black body, it would emit power at a rate of 110 watts. What is the emissivity of the object?

Here, since $H_B = R_B$

$$\Rightarrow e = \frac{R}{R_B} = \frac{40 watts (area\ of\ object)}{110 watts (area\ of\ object)}$$

so that e = 0.364

☐☐ Explain why for a storm window there is more heat transferred if the air gap between the glass of the window and the glass of the storm window is large.

The reason is that the convection currents caused by the heating of the air next to the hot side with its resultant rise, and the cooling of the air next to the cold side with the resultant drop causes heat to be transferred by convection. For this reason, the space between the panes of glass is usually fairly small to eliminate these convection currents.

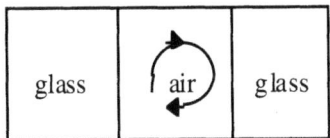

☐☐ How large should a solar collector be in order to supply 20 kw of power if the irradiance is 450 watt/m² and the collector has efficiency of 50%? Ignore the radiant emittance of the solar collector.

$$H = \frac{energy}{time\ area} \Rightarrow HA = \frac{energy}{time}$$

Thus, 0.5(450 watt/m²)A = 20,000 watts. Hence, the area of the solar collector should be 88.9 m².

☐☐ Find the coefficient of thermal conductivity of a material if 42 J/s flows through a slab 2 mm thick and 55 cm² is the area if the temperature on one side is 99 °C and that on the other is 35 °C.

$$H = \frac{kA(T_h - T_c)}{L} \Rightarrow \frac{HL}{A(T_h - T_c)} = k = \frac{(42 J/s)(.002m)}{(55 \times 10^{-4} m^2)(64°C)} = 0.239\ watt/m°C$$

☐☐ Why aren't buildings insulated with air? Air has a lower coefficient of thermal conductivity than does fiber glass, a commonly used insulating material, so all other things being equal air would conduct less heat than fiber glass.

If air alone were used, on windy days the air would be moving and thus there would be heat transfer by convection currents. Thus, by using fiberglass rather than air, heat transfer by convection is greatly reduced.

□□ A brass bar has its sides insulated and one end is in a steam bath at 100 °C and the other end is in a steam bath at 0 °C. If the length of the bar is 30 cm and its cross section is 3 cm², find the amount of ice melted in 10 minutes. (Coefficient of thermal conductivity of brass is 100 watt/m°C). (Neglect radiation losses; heat of fusion of water is 335 J/g).

Here, $Q = \dfrac{kA(T_h - T_c)time}{L} = \dfrac{(100 J/s - m°C)(3x10^{-4} m^2)(100°C)(600s)}{.3m}$

$Q = 6000J = mL_f \Rightarrow \dfrac{6000J}{335 J/g} = m = 17.9g$

or .0179 kg of ice melted in 10 minutes.

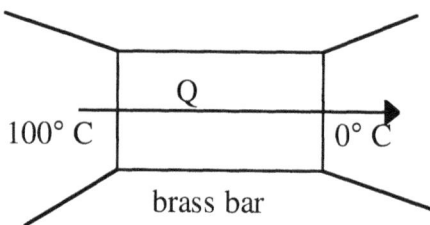

□□ A flat plate is maintained at 90 °C, and the air on both sides is at 15 °C and atmospheric pressure. Find the heat lost by natural convection per m² in one hour if (a) the plate is vertical, and (b) the plate is horizontal (both sides). The convection coefficient for a vertical plate is 1.77 $(\Delta T)^{1/4}$ (J/sm² °C) and for a horizontal plate facing down is 1.31$(\Delta T)^{1/4}$ (J/sm² °C) and for a horizontal plate facing up is 2.49$(\Delta T)^{1/4}$ (J/sm² °C).

(a) Here H = hAΔT, $\dfrac{Q}{t} = hA\Delta T \Rightarrow \dfrac{Q}{A} = h(\Delta T)t$

$\dfrac{Q}{A} = [1.77(75)^{\frac{1}{4}} J/s - m^2 °C](75°C)3600s$ for one side.

$\dfrac{Q}{A} = 1.41x10^6 J/m^2$ for one side, hence $\dfrac{Q}{A} = 2.82x10^6 J/m^2$ for both sides.

(b) If the plate is horizontal, calculate the heat loss from the bottom and top sides separately. Thus, H = hAΔT = Q/t ⇒ Q/A = h(ΔT)t. So for bottom side,

$\dfrac{Q}{A} = (1.31(75)^{1/4} J/s - m^2°C)(75°C)(3600s)$

$\dfrac{Q}{A}_{bottom} = 1.04x10^6 J/m^2$

for top side,

$\dfrac{Q}{A} = 2.49(75)^{1/4} J/s - m^2°C(75°C)(3600s)$

$\dfrac{Q}{A}_{top} = 1.98x10^6 J/m^2$

Hence, if the plate were horizontal, the heat lost in one hour per square meter is 3.02 x 10⁶J/m².

☐☐ A wood door of dimensions 0.9 m by 2 m by 4 cm thick (coefficient of thermal conductivity of 0.04 watt/m °C) has a hot side temperature of 23° C and a cold side temperature of -10°C. Find (a) the heat current through the door and (b) by what factor the heat current is increased if a glass window of dimensions 0.4 m by 0.4 m by 2 mm thick is placed in the door. (Coefficient of thermal conductivity of glass is 0.8 watt/m° C).

(a) $H = \dfrac{kA(T_h - T_c)}{L} = \dfrac{(.04 watt/m°C)(1.8m^2)(33°C)}{.04m}$

$H = 59.4$ J/s

(b) with the window, $H = H_{wood} + H_{glass}$

$H = \dfrac{(.04 watt/m°C)(1.64m^2)(33°C)}{.04m} + \dfrac{(.8 watt/m°C)(.16m^2)(33°C)}{.002m}$

$= 54.1$ J/s $+ 2110$ J/s

thus, with window, $H \cong 2160$
without window, $H \cong 59.4$ J

So the heat current with the window has been increased by a factor of 36.4 times.

☐☐ A vertical pipe (diameter 8 cm, height 5 m) carries a fluid at 150° C while the outside temperature is at 23° C. If the convection coefficient for a vertical pipe is 1.32 $(\Delta T/D)^{1/4}$ where D is the diameter of the pipe. Find (a) the heat current for convective transfer, and (b) the heat transferred by convection in 30 minutes. (The area of a cylinder is $2\pi rh$).

(a) $H = hA\Delta T$

$H = \left[1.32\left(\dfrac{127}{.08}\right)^{1/4} watt/m^2°C\right](6.28)(.04m)(5m)127°C$

$H \approx 1330$ J/S

(b) $H = Q/T \Rightarrow Q = HT = (1330$ J/S$)(1800$S$)$
$Q = 2.39 \times 10^6$ J Transferred in 30 minutes.

☐☐ A blackened cylindrical container (diameter .06 m, height 0.15 m, emissivity of 1) is used to store liquid helium at 4.2 K, while the can is surrounded by liquid nitrogen at 77 K. Assume that there is a vacuum between the can and the liquid nitrogen (only radiative transfer can occur). Find the heat current transferred to the can by radiation.

$H = H - R_B$ (since the can is gaining energy)

$= Ae\sigma T_0^4 - Ae\sigma T^4$ (T_0 = temperature of surround)

$= Ae\sigma(T_0^4 - T^4)$ (T = temperature of can)

$H = (2\pi rh)(\sigma)(T_0^4 - T^4)$

$H = (6.28)(.03m)(.15m)(56.7 \times 10^{-9} watt/m^2 K^4)\left[(77J)^4 - (4.2K)^4\right]$

$H = .056$ J/s

❑❑ In the previous problem, if the helium is at its boiling point, and the heat of vaporization of helium is 20 J/g, how many grams of helium are vaporized in one hour?

$Q = mL_v$, and $H = Q/t \Rightarrow Q = (.056 \text{ J/s})(3600\text{s})$

$Q = 203 \text{ J}$

Thus, $Q/L_v = m = \dfrac{203 J}{20 J/g} = 10.2$ g of helium vaporized.

THERMODYNAMICS

THERMODYNAMICS HAS THE FOLLOWING IDEAS:

- A heat reservoir is a body of such large mass that it can absorb or reject unlimited quantities of heat without any appreciable change in temperature.

- An adiabatic process is a process in which no heat is allowed to enter or leave a system.

- An isothermal process is a process in which the temperature of a system doesn't change.

- An isobaric process is a process in which the pressure remains constant.

- An isochoric process is a process in which the volume remains constant.

- The work done on an hydrostatic system (a system that can be specified by the volume V, pressure P, and temperature T) is given by PΔV if the process is isobaric.

- The first law of thermodynamics says that $\Delta U = Q - W$, where U is the internal energy of the system, Q is heat, and W is work done by the system.

- For an ideal gas, the work done for an isothermal process is given by $W = nRT\ln\left(\dfrac{V_f}{V_i}\right)$, where n is number of moles of gas and R is the universal gas constant, T is absolute temperature, and V is volume.

- For a monatomic ideal gas, the work done during an adiabatic process is given by $W = (3/2) nR(T_i - T_f)$.

- For an adiabatic process, $P_i V_i^\gamma = P_f V_f^\gamma$ when $\gamma = \dfrac{C_P}{C_V}$, where C_P = heat capacity at constant pressure, and C_V = heat capactiy at constant volume.

- For any ideal gas $C_P - C_V = nR$, whether the gas is monatomic or diatomic.

- The internal energy of an ideal gas is a function of temperature, i.e., $\Delta U = 0$ for an ideal gas. That undergoes an isothermal process.

- A heat engine consists of a system that undergoes the same cycle over and over again. The engine absorbs heat from a hot reservoir, delivers work to the surround, and rejects leftover heat to a cold reservoir.

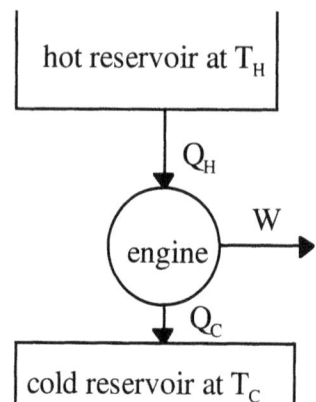

- The efficiency of a heat engine is given by $\frac{work\ done}{heat\ supplied}$, i.e. eff = $\frac{W}{Q_H} = \frac{Q_H - Q_C}{Q_H}$

- Eff = $1 - \frac{Q_C}{Q_H}$

- A reversible process is a process in which both the system and the surround can be returned to their original states before the process occurred.

- An irreversible process is a process in which the system and/or the surround can not be returned to their original states before the process occurred.

- A Carnot engine is an idealized heat engine that operates reversibly, while no real engine does. no heat engine can be more efficient than a Carnot engine, and the efficiency of a Carnot engine is given by eff = $1 - \frac{Q_C}{Q_H} = 1 - \frac{T_C}{T_H}$ where T_C and T_H are the temperature of the hot and cold reservoirs in kelvins.

- A refrigerator is a system that performs the same cycle over and over again. Work is done on the system, which extracts heat from a cold reservoir, and rejects heat to a hot reservoir.

- The coefficient of Performance (C.O.P.) of a refrigerator or air conditioner is given by
C.O.P. = $\frac{Q_C}{W}$

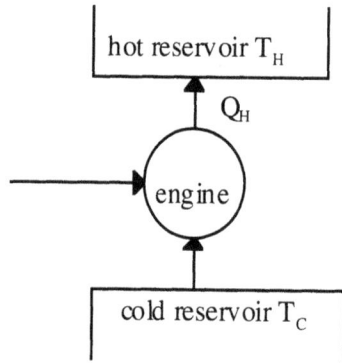

- There are two forms to the Second law of Thermodynamics, which are equivalent:
    1. The Kelvin-Planck form: there exists no process whose sole result is the absorption of heat from a reservoir and the conversion of this heat to work.

2. The Clausius form: there exists no process whose sole result is the transfer of heat from a cooler to a hotter body.

- Entropy (symbol S) is a measure of the order of a system. Increased entropy means a greater degree of disorder and decreased entropy a lesser degree of disorder. Entropy is a measure of the energy that is unavailable to do work because of an irreversible process.

- The entropy principle says that $(\Delta S)_{Universe} \geq 0$, where $\Delta S = 0$ if a reversible process takes place, and $\Delta S > 0$ if an irreversible process takes place.

SOLVED PROBLEMS

☐☐ **A system does 250 Joules of work and its internal energy decreases by 680 J. How much heat was accepted or rejected by the system?**

$\Delta U = Q - W$
$Q = \Delta U + W$
$Q = -680 J + 250 J$
$Q = -430 J$
Thus, the system rejected 430 J of heat [Q - $\Rightarrow$ heat leaves system Q + $\Rightarrow$ heat enters system].

☐☐ **A heat engine absorbs 300 J of heat from a hot reservoir at 400 K, delivers 100 J of work to the surround, and rejects the leftover heat to a cold reservoir at 200 K during one cycle. Find (a) how much heat was rejected to the cold reservoir and (b) the efficiency of the heat engine.**

(a) $Q_H = W + Q_C \Rightarrow Q_H - W = Q_C = 300 J - 100 J == 200 J$ is the heat rejected to cold reservoir.

(b) Eff = $\dfrac{W}{Q_H} = \dfrac{100 J}{300 J} = \dfrac{1}{3}$ or 33%

☐☐ **For the previous problem, what would be the efficiency of a Carnot engine operating between the same two reservoirs?**

Efficiency of a Carnot engine = $1 - \dfrac{T_C}{T_H} = 1 - \dfrac{200 K}{400 K}$

Eff = 1/2 or 50%
Note that no engine can be more efficient than a Carnot engine.

☐☐ **A monatomic ideal gas ($\gamma = 5/3$) is compressed adiabatically and its volume is halved. Determine by what factor the pressure is increased.**

For an adiabat, $P_1 V_1^\gamma = P_2 V_2^\gamma$

Thus, $P_2 = \dfrac{P_1 V_1^\gamma}{V_2^\gamma} = P_1 \left(\dfrac{V_1}{\dfrac{V_1}{2}} \right)^{5/3} = P_1(2^{5/3})$ so that $P_2 = 3.17 P_1$ or P_2 is 3.17 times greater than P_1.

☐☐ **If the coefficient of performance of a refrigerator is 5.2, how much work is done in removing 4500 J of heat from the inside of the refrigerator?**

C.O.P. = $\frac{Q_C}{W} \Rightarrow W = \frac{Q_C}{C.O.P.} = \frac{4500J}{5.2}$. Thus, the work done is 865 J.

☐☐ **An isothermal process causes 2 moles of an ideal gas to expand from 0.1 m³ to 0.4 m³. What work is done by the system if the temperature is 27 °C?**

Work for an isothermal process is given by nRT ln $\frac{V_f}{V_i}$

Thus, work = (2moles)(8.314 J/mol-K)(300K)ln$\left[\frac{.4m^3}{.1m^3}\right]$.

Thus, work = 6920 J. Note that temperature here is absolute temperature.

☐☐ **0.45 moles of helium expands isothermally at 350 Kelvins and does 1500 J of work. Find the ratio of the final volume to the initial volume of the helium.**

For an isothermal process, work = nRT ln $\frac{V_f}{V_i}$

Thus, 1500 J = (0.45 moles)(8.314J/mol-K))(350 K)ln $\frac{V_f}{V_i}$

1.15 = ln $\frac{V_f}{V_i}$, thus, $\frac{V_f}{V_i} = e^{1.15} = 3.14$

so that the final volume is 3.14 times the initial value.

☐☐ **5 moles of a monatomic gas is compressed isothermally at 325 K. What is the final temperature of the gas?**

325 K

☐☐ **If for the previous problem, 500 J of work are done on the gas, what is the heat removed?**

Here, from the First Law of Thermodynamics, ΔU = Q - W. But if a process is isothermal, the internal energy doesn't change. Thus, 0 = Q - W ⇒ Q = W = 500J.

☐☐ **An egg falls to the floor. Has the order of the universe increased?**

(ΔS)$_{Universe}$ ≥ 0. For an irreversible process (ΔS) > 0. Thus, the order of the universe has decreased (or the disorder of the universe has increased).

☐☐ **During an isobaric process, a gas undergoes an expansion from 0.01m³ to 0.2 m³ at atmospheric pressure. What work was done by the gas? (1 atm = 1.01 x 10⁵ N/m²)**

For an isobaric process, work = PΔV
work = (1.01 x 10⁵ N/m²)(0.2m³ - 0.01m³)
work done = 19,200 J

☐☐ 1.6 moles of an ideal monatomic gas undergoes an adiabatic compression. The initial temperature is 23 ° C, and the final temperature is 200 ° C. What work is done?

For an ideal monatomic gas, w = (3/2)nR(T_i - T_f)
                 w = (3/2) (1.6 moles)(8.314 J/mol-K)(300 K - 473 K)
                 work = -3450 J

Note that the – means that work was done on the gas.

☐☐ An adiabatic process is an idealization. What could be done to approximate an adiabatic process?

If the process is done quickly, so that heat doesn't have time to enter or leave the system the process will be approximately adiabatic.

☐☐ Consider the figure. An engine extracts 500 J of heat from the hot reservoir and converts the heat to work with no heat rejected to the cold reservoir. Why is this impossible?

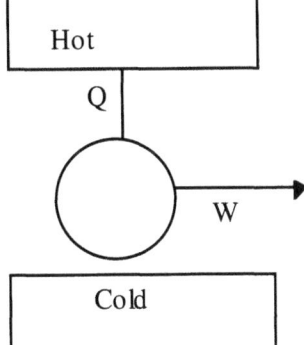

Because it violates the Kelvin-Planck statement of the Second Law of Thermodynamics, which states that there exists no process whose sole result is the absorption of heat from a hot reservoir and the conversion of this heat to work.

☐☐ A freezer converts 0.6 kg of water at 10 ° C to ice at -10 ° C. If the coefficient of Performance of the freezer is 4.0, what is the electrical energy required? (Specific heat of water is 4186 J/kg °C, the specific heat of ice is 2100 J/kg °C, and the heat of fusion of water is 335 x 10^3 J/kg.)

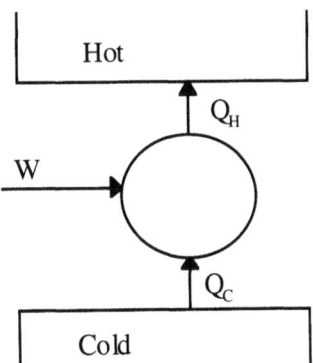

Here, the definition of the coefficient of Performance is C.O.P. = $\dfrac{Q_C}{W}$

so that work needed = $\dfrac{Q_C}{C.O.P.}$

to find Q_C: $Q_C = mc\Delta t_{water} + mL + mc\Delta t_{ice}$

$= (0.6 kg)(4186 J/kg°C)(10°C) + (0.6kg)(335 \times 10^3 J/kg) + (0.6kg)(2100 J/kg°C)(10°C)$
$Q_C = 2.39 \times 10^5$ J

Thus, work required = $(2.39 \times 10^5 J)/4 \approx 59,700$ J
Since, ΔEnergy = work done, 59,700 J of electrical energy is required.

☐☐ **Consider the figure. Why is this impossible?**

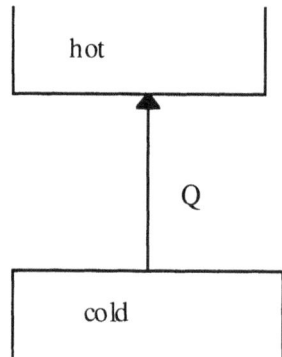

This is impossible because it violates the Clausius statement of the Second Law of Thermodynamics, which says that there exists no process whose sole result is the transfer of heat from a cooler to a hotter body.

☐☐ **Assuming that two moles of air are heated from 1 to 100° C at one atmosphere, find (a) the heat added, (b) the work done on the gas, and (c) the change in the internal energy of the air. (M_{air} = 29g/mole, and assume the specific heat of air at room temperature is 1000J/kg °C. Assume air acts as an ideal gas.)**

(a) Here, $Q = mc\Delta T$
$= (.058$ kg$)(1000$J/kg °C$)(100°$ C$)$
$Q = 5800$ J

(b) For a constant pressure process of an ideal gas, work = $P\Delta V$. But since for an ideal gas,
$P_2V_2 = nRT_2$
$P_1V_1 = nRT_1$, and $P_2 = P_1$

Then by subtraction, $P_1(V_2 - V_1) = nR(T_2 - T_1)$ so that work = $P\Delta V = nR\Delta T$ in a constant pressure process. Thus, the work done on the gas is given by
work = (2 moles)(8.314J/mol-K)(373K - 273 K) = 1660 J.

(c) Since the First Law of Thermodynamics says that $\Delta U = Q - W$, the change in internal energy of the air is given by $\Delta U = 5800$ J - 1660 J = 4140J

☐☐ **Six moles of an ideal gas undergoes the cycle A → B → C → A as in the diagram. The portion of the cycle from C → A is an adiabatic process. (a) What kind of process is the process A → B, (b) what kind of process is the process from B → C, and (c) what is the work done on the system when it goes form A → B, and (d) how much heat enters the system when it goes from C → A?**

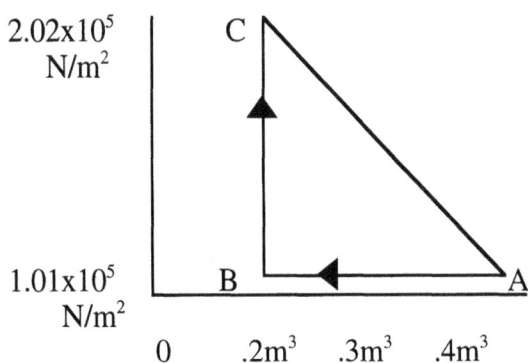

(a) The process from A → B is carried out at constant pressure, thus it's an isobaric process.

(b) The process from B → C is carried out at constant volume, hence it is an isochoric process.

(c) For an isobaric process, work = $P\Delta V$ = $(1.01 \times 10^5 \text{ N/m}^2)(.2\text{m}^3 - .4\text{ m}^3)$
work = $-20{,}200$ J which is work done on the gas.

(d) Since the process form C → A is adiabatic, no heat enters or leaves the system.

☐☐ For the previous problem, assume that the temperature is at 700K at point A. What is the temperature at point B:

$P_B V_B = nRT_B$
$P_A V_A = nRT_A$

Subtracting these equations leaves:
$P(V_B - V_A) = nR(T_B - T_A)$ or $P\Delta V = nR\Delta T$ for a constant pressure process. Hence, from the previous problem, $-20{,}200$ J = (6 moles) (8.314 J/mol-K)ΔT
Thus, $\Delta T = -405$ K, since $\Delta T = T_B - T_A$. The temperature $T_B = -405$ K + 700 K or $T_B = 295$ K.

☐☐ A refrigerator (coefficient of Performance 3.2) is used to change 2 kg of water at 20 °C to ice at 0 °C. Find (a) the amount of heat that is taken from the water and rejected to the hot reservoir, and (b) how much electrical energy is required to accomplish this task. (The heat of fusion of water is 335 x 10^3 J/kg, and the specific heat of water is 4186 J/kg °C).

(a) to change the water at 20°C to ice at 0 °C, $Q = mc\Delta T| + mL_f$
$Q = 2$ kg(4186 J/kg °C)(20 °C) + 2 kg(335 x 10^3 J/kg)
$Q = 8.37 \times 10^5$ J

(b) For a refrigerator, C.O.P. = $\dfrac{Q_C}{W}$

Thus, $W = \dfrac{Q_C}{C.O.P.} = 2.62 \times 10^5 J$

Since change in energy equals work done, the electrical energy required for this task is 2.62 x 10^5 J.

☐☐ For the previous problem, how long would a 2400 watt heater have to run to put in a kitchen the same amount of heat as the refrigerator does when it cools and then freezes the water?

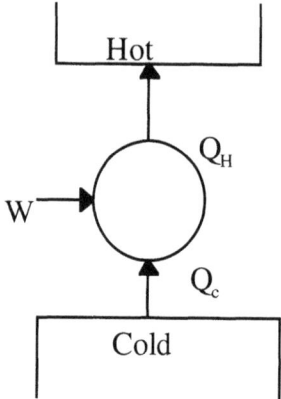

Here, $Q_H = W + Q_C = 1.1 \times 10^6$ J

Since Power $= \dfrac{\Delta E}{time} \Rightarrow \dfrac{\Delta E}{Power} = \dfrac{1.1 \times 10^6 J}{2400 J/s} = $ time

time = 457 s or 7.63 minutes.

☐☐ Given that the molar heat capacities for an ideal monatomic gas are (5/2)R for constant pressure and (3/2) R for constant volume, one mole of an ideal monatomic gas is carried around the cycle (a →b→c→d→a) shown in the figure. Find (in terms of P_0 and V_0): (a) the heat that enters the system, $Q_1 + Q_2$, (b) the heat $Q_3 + Q_4$, and (c) the efficiency of the cycle as a heat engine.

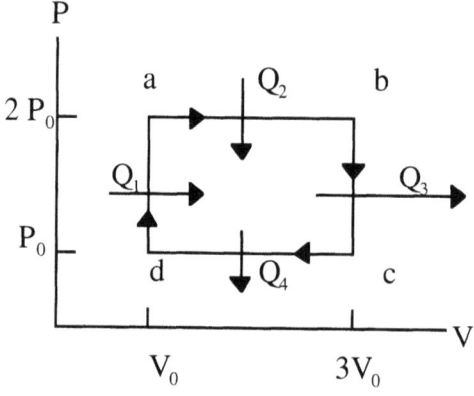

(a) For a constant volume process $Q = nc_V \Delta T$, and for a constant pressure process $Q = nc_P \Delta T$. Note that as has been shown above, $P\Delta V = nR \Delta T$ for a constant pressure process and $P_2V_2 = nRT_2$ and $P_1V_1 = nRT_1$
Then, $P_2V_2 - P_1V_1 = nR(T_2 - T_1)$ or $V\Delta P = nR\Delta T$ for a constant volume process.

Thus, $Q_1 = nc_V \Delta T = (nc_V)\left(\dfrac{V\Delta P}{nR}\right) = \dfrac{3}{2}R\left(\dfrac{1}{R}\right)(V_0)(2P_0 - P_0)$

$Q_1 = (3/2)V_0 P_0$

and $Q_2 = nc_P \Delta T = (nc_P)\left(\dfrac{P\Delta V}{nR}\right) = \dfrac{n\left(\dfrac{5}{2}\right)R(2P_o)(3V_0 - V_0)}{nR}$

$Q_2 = 10P_0V_0$

Hence, the heat that enters the system is $11.5\ P_0V_0$.

(b) To find the heat that leaves the system,

$$Q_3 = nc_V\Delta T = n\left(\frac{3}{2}R\right)\left(\frac{V\Delta P}{nR}\right)$$

$$= \frac{3}{2}(3V_0)(P_0 - 2P_0)$$

$$= -\frac{9}{2}P_0V_0$$

$$Q_4 = nc_P\Delta T = \frac{n\left(\frac{5}{2}R\right)(P_0)(V_0 - 3V_0)}{nR}$$

$$= \frac{5}{2}P_0(-2V_0) = -5P_0V_0$$

Thus, the heat that leaves the system is $9.5\ P_0V_0$.

(c) Since the efficiency of a heat engine is given by eff $= \dfrac{W}{Q_H} = 1 - \dfrac{Q_C}{Q_H}$, the efficiency of this system as a heat engine is given by $1 - \dfrac{9.5P_0V_0}{11.5P_0V_0}$

and eff = .174 or 17.4%.

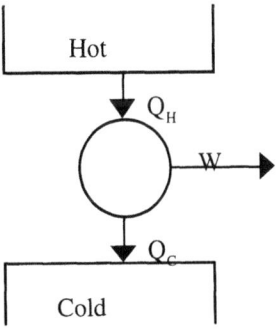

☐☐ **If molar heat capacity for a gas at constant volume is (5/2)R, what is the molar heat capacity at constant pressure?**

$C_P - C_V = nR$ or $C_P - C_V = R$. Thus, $C_P = R + C_V = R + (5/2)R \Rightarrow C_P = (7/2)R$, so that molar heat capacity at constant pressure is $(7/2)R$.

☐☐ **In nature, people age, eggs rot, etc. Are these reversible processes? Does the entropy of the universe increase or remain the same?**

The processes mentioned above are irreversible processes. The entropy principle says that the change in entropy (ΔS) of the universe increases because of any irreversible process. Another way of saying this is that the order of the universe has decreased (or that the disorder of the universe has increased) as a result of these processes.

❑❑ **An inventor has claimed to have perfected a process that reaches a temperature of absolute zero (0 K). Would this be feasible?**

Not without violating the Third Law of Thermodynamics, which says that it is impossible to lower the temperature of any system to absolute zero in a finite number of steps.

❑❑ **A refrigerator absorbs 800 J of heat per cycle and has a coefficient of Performance of 4.5. Find (a) how much work this requires and (b) how much heat is rejected to the surround.**

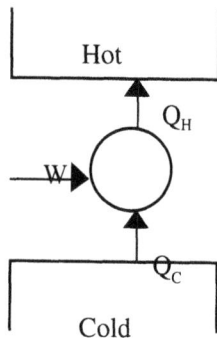

For a refrigerator, $C.O.P. = \dfrac{Q_C}{W}$

(a) Thus, $W = \dfrac{Q_C}{C.O.P.} = \dfrac{800J}{4.5} = 178J$

(b) The heat rejected can be found because $Q_H = W + Q_C = 978$ J.

WAVES AND SOUND
THE MOTION OF WAVES

THE FUNDAMENTAL IDEAS ARE THE FOLLOWING:

- Wave motion transfers energy from one point to another.

- Waves are classified in terms of how the individual particles that make up the wave move with respect to the direction of wave travel (propagation).

- A transverse wave is a wave in which the particles that make up the wave vibrate back and forth perpendicularly to the direction of wave travel.

- A longitudinal wave is a wave in which the particles that make up the wave vibrate back and forth along the direction in which the wave travels.

- A representation of a transverse wave:

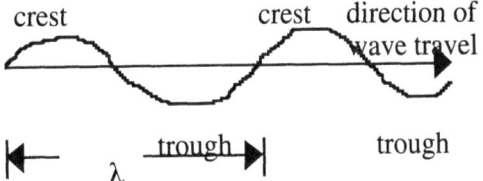

- A representation of a longitudinal wave:

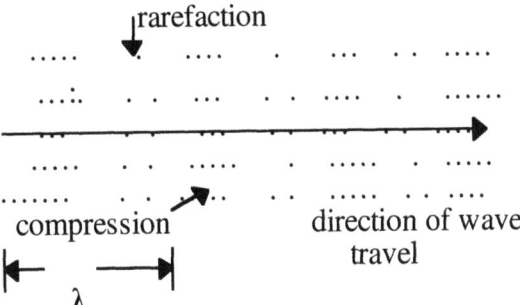

- The waves that spread out from the impact of a rock in a pond and light waves are examples of transverse waves.

- Sound waves are an example of a longitudinal wave.

- Waves are repetitive, and the distance the wave travels before it repeats itself is called the wavelength, λ.

- The wave equation says that $v = f\lambda$, where v is the velocity of the wave, f is the frequency, and λ is the wavelength.

- The energy transmitted per unit time through a unit area perpendicular to the direction of wave travel is called the intensity of the wave motion.

- The intensity is proportional to the square of the frequency and the square of the amplitude of the vibrations of the particles that make up the wave.

- For a spherical wave, the intensity varies inversely as the square of the distance from the source if there is no energy loss. If the energy cannot spread out freely in all directions, the intensity doesn't vary inversely as the square of the distance.

- Huygen's principle says that every point on a wave front acts as a new source sending out secondary waves.

- For reflected waves, there are two laws:

  (a) The angle of incidence equals the angle of reflection.

  (b) The incident ray, the reflected ray and the normal to the reflecting surface all lies in the same plane.

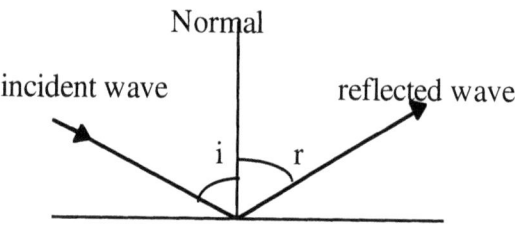

Note that "i" is angle of incidence, and r is angle of reflection, and that both "i" and r are measured with respect to the normal.

- The bending of waves as they pass from one medium to another is called refraction.

- If V_1 is the velocity of the wave in medium 1, and V_2 is the velocity of the wave in medium 2, then $n = \text{constant} = \dfrac{\sin i}{\sin r} = \dfrac{V_1}{V_2}$, where n is called the index of refraction of medium 2 relative to medium 1. The incident ray, the refracted ray and the normal to the surface all lie in the same plane.

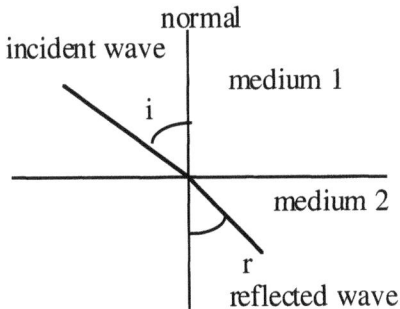

- When waves travel from one medium to another the frequency of the wave does not change.

- The superposition principle says that when two or more waves pass a given point at the same instant, the displacement of the resultant wave is the sum of the individual displacements.

- The addition of two or more waves is called interference.

- If the displacement of the resultant wave is twice the displacement of the individual wave, then constructive interference is said to have occurred.

- If the displacement of the resultant wave is zero, then the waves are said to exhibit destructive interference.

- Two waves of different amplitudes may interfere; the resultant amplitude is the sum of the individual amplitudes.

- A standing wave is the result of interference between two waves that are travelling in opposite directions.

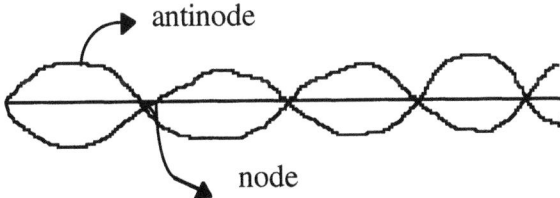

- Standing waves may result from any kind of wave motion.

SOLVED PROBLEMS

❏❏ **A piano emits a sound that has a frequency of 425 Hz. Find (a) the period of the vibration, and (b) the wavelength of the sound. (Assume V = 340 m/s)**

(a) $V = f\lambda$
340 m/s = (425 cy/s)λ, λ = 0.8 m
Since for Harmonic motion T = 1/f, here, T = 0.0024 s.

❏❏ It is 8 m from crest to crest in a wave train. If 26 waves pass a given point in a minute, find (a) the wavelength of the wave, (b) the frequency of the waves, and (c) the speed of the waves.

(a) Since the distance that a wave travels before it repeats itself is the wavelength λ, here $\lambda = 8$ m.

(b) Frequency is $\dfrac{\text{\# of cycles}}{\text{time}}$, thus, $f = \dfrac{26 \text{ waves}}{60 s} \Rightarrow f = .433 Hz$

(c) $V = f\lambda = (.433 \text{ Hz})(8 \text{ m}) = 3.47$ m/s.

❏❏ A wave of wavelength 16 m is travelling through a distance of 45 m. How many waves are there in the 45 m distance?

Since one wave occupies 0.16 m, the # of waves is $\dfrac{45 m}{.16 m/wave}$ or 281 waves.

A general rule says that # of waves = $\dfrac{\text{distance}}{\text{wavelength}}$

❏❏ A swimmer is under water when the race starter fires her gun. The sound wave has an angle of incidence of 13°. Find (a) the index of refraction for a sound wave going from air to water and (bd) the angle of refraction of the sound wave. (Speed of sound in air is 340m/s, in water 1450m/s)

(a) Index of refraction = $\dfrac{V_1}{V_2}$

$n = \dfrac{340 m/s}{1450 m/s} = .234$

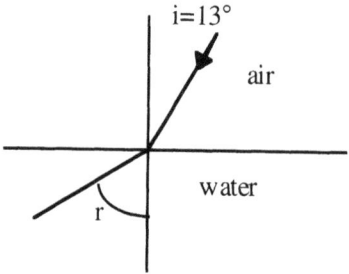

(b) Since for refraction $n = \dfrac{\sin i}{\sin r} \Rightarrow \sin r = \dfrac{\sin i}{n} = \dfrac{\sin 13°}{.234} = .961$

$r = 74°$

☐☐ A submarine sends out a sonar pulse, and 5 s later a reflection returns. If the speed of sound in seawater is 1540 m/s, how far away is the ship?

$2x = vt \Rightarrow 2x = (1540 \text{ m/s})(5 \text{ s}) = 7700 \text{ m}$
$x = 3850 \text{ m}$ so the ship is 3850 m from the sub.

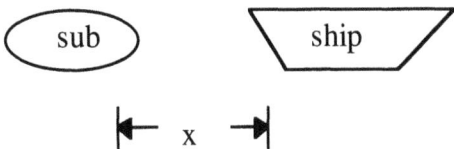

☐☐ Standing waves are produced in a stretched rope as in the figure. The distance between nodes is 0.762 m. Find (a) the wavelength, and (b) the frequency of the standing wave if the waves that make up the standing wave have a velocity of 10.7 m/s.

(a) Note that the distance between two nodes is one half wavelength. Thus, here $\lambda = 1.52$ m.

(b) Since $V = f\lambda \Rightarrow f = \dfrac{V}{\lambda} = \dfrac{10.7 m/s}{1.52 m}$

$f = 7.02$ Hz

☐☐ A frequency of 180 Hz produces standing waves with 30 cm between adjacent nodes. Find (a) the wavelength of the waves, and (b) the speed of the waves.

(a) Here, $\lambda = 2(.3,) = .6$ m
(b) $V = f\lambda = (180 \text{ cy/s})(.6 \text{ m}) = 108$ m/s

☐☐ A wave has a wavelength of 50 cm in air. Find its wavelength in water. (Assume speed of sound in air is 340m/s, in water 1450m/s)

Since $n = \dfrac{V_1}{V_2}$, then $\dfrac{V_1}{V_2} = \dfrac{f_1 \lambda_1}{f_2 \lambda_2}$

Thus, since $f_1 = f_2$, $\lambda_2 = \lambda_1 \cdot \dfrac{V_2}{V_1} = (.5m)\left(\dfrac{1450 m/s}{340 m/s}\right)$

Hence, the wavelength of the wave in water is 2.13 m.

☐☐ An underwater explosion produces a sound wave that approaches the surface of the water with an angle of incidence of 60°. Find the angle of refraction. (Assume speed of sound in air is 340m/s, in water 1450m/s)

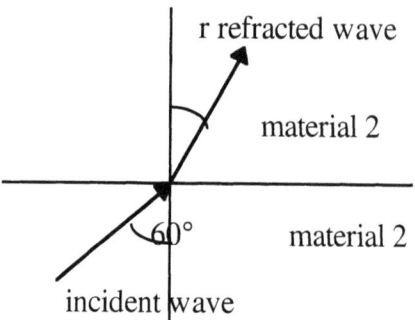

$$n = \frac{\sin i}{\sin r} = \frac{V_1}{V_2}, \text{ thus } \frac{\sin 60°}{\sin r} = \frac{1450 m/s}{340 m/s} \text{ and hence,}$$

$$\sin r = \sin 60° \left(\frac{340 m/s}{1450 m/s}\right) = .203 \text{ and } r = 11.7°$$

☐☐ A tuning fork vibrates at 440Hz. How many times does the tuning fork vibrate in the time required for the sound from the tuning fork to travel 10 m?

Since $f = \frac{1}{T} \Rightarrow T = \frac{1}{f} = \frac{1}{440} s$, then 1 vibration = $\frac{1}{440} s$

The time for the sound to travel 10 m can be found by $x = vt \Rightarrow 10 m = (340 m/s)t$ or

$t = .029$ s. Since 1 vibration = $\frac{1}{440} s \Rightarrow .029 s \left(\frac{1 vibration}{\frac{1}{440} s}\right) = 12.9 vibrations$

The tuning fork vibrates 12.9 times in .029 s.

☐☐ A stone is thrown into a pond, and circular ripples (transverse waves) result. If at a certain instant the first crest is 5 m from the place where the stone hit the water, and the seventh crest is 50 cm past the point of impact, what is the wavelength of the waves?

Here 6 waves = 4.5 m. Thus, the distance from crest-to-crest is $\frac{4.5m}{6} = 0.75m$.

Hence, $\lambda = 0.75$ m.

☐☐ A sonar pulse (frequency 35,000 Hz) is generated at a place where the speed of sound is 1540 m/s. If the echo from a submarine takes 5 seconds to return to the sonar receiver, find (a) the wavelength of the sonar wave, and (b) how far away is the submarine from the sonar transmitter?

(a) $V = f\lambda \Rightarrow 1540$ m/s = $(35,000$ cy/s$)\lambda$
$\lambda = 0.044$ m

(b) Since $x = vt \Rightarrow x = (1540$ m/s$)(5$ s$) = 7700$ m, the sonar pulse travels to the sub a and back again, hence the sub is (x/2) or 3850 m away.

◻◻ There is an old saying that during a thunderstorm, one counts the seconds between the lightning flash and the clap of thunder to get an approximate idea of how far away the lightning strike was; five seconds means the strike was a mile away, ten seconds means the strike was two miles away, etc. Is this a reasonable saying?

Since one mile is 1609 m, and sound travels at about 340 m/s:
x = vt
x = (340 m/s)(5 S) = 1720 m.
Thus, it is not perfect, but it is an approximate way to find out how far away a lightning strike was.

◻◻ If 60 waves with a speed of 245 m/s pass a given point each second, find (a) the wavelength, and (b) the time it takes for one wave to pass the point.

(a) Since frequency = $\frac{\# \text{ of waves}}{\text{time}}$ ⇒ $f = \frac{60 \text{ waves}}{s}$

Hence, V = fλ or 245 m/s = (60 cy/s)λ.
λ = 4.08 m

(b) If there are 60 waves past a point in one second that means that it takes $\frac{1}{60}$ s for one wave to pass the point.

SOUND

THE FUNDAMENTAL IDEAS ARE:

- Sound is a longitudinal wave.

- The velocity of sound through any medium depends on the density and elastic properties of the medium.

- In a wire or rod, the velocity of sound is given by $V = \sqrt{\dfrac{Y}{\rho}}$, where Y is the Young's modulus and ρ is the density of the material.

- In a fluid, $V = \sqrt{\dfrac{B}{\rho}}$, where B is the bulk modulus and ρ is the density.

- For a gas, B is the adiabatic bulk modulus, given by γP, where γ is a ratio of C_P/C_V, and P is the pressure. For a monatomic gas, $\gamma = 1.67$, and for a diatomic gas γ is 1.4. Thus, for any gas $V = \sqrt{\dfrac{\gamma P}{\rho}}$.

- For an ideal gas $PV = nRT$, hence the velocity of sound in a gas can be written as

- $V = \sqrt{\dfrac{\gamma RT}{M}}$, where M = molecular mass, T is absolute temperature and R is the universal gas constant.

- The speed of sound in a gas can be found by $V = V_0 \sqrt{\dfrac{T}{273}} = V_0 \sqrt{1 + \dfrac{t}{273}}$, where V_0 is the speed at 0° C and t is the temperature in Celsius degree.

- The range of frequencies that can be heard by the human ear is from 20 to 20,000 Hz, although this varies widely with the individual. Ultrasonic sound has frequencies above 20,000 Hz.

- To a listener, pitch, loudness and quality characterize sound. Pitch depends on the frequency of the fundamental (lowest frequency present in the sound wave). Loudness is the magnitude of the sensation produced in the ear by the sound. Quality depends on the source, for example, a sound produced by a clarinet has a different quality than a sound produced by a piano.

   The intensity of a sound is measure in decibels. The threshold of hearing has an intensity level of $I_0 = 1 \times 10^{-12}$ watt/m^2, and the intensity level β (in decibels) is found by

   β (in decibels) $= 10 \log \left(\dfrac{I}{I_0}\right)$. Note that β is dimensionless, and that when I = the threshold of hearing I_0, $\beta = 0$ decibels.

- If two sources of sound have frequencies that are slightly different, the intensity of the sound fluctuates. The fluctuations in intensity are called beats. The number of beats per second is given by $n = f_1 - f_2$.

- The doppler effect is the change in pitch or frequency of the sound detected by an observer because the sound source and the observer have different velocities relative to the material through which the sound is moving.

A general case to find the observed frequency $f_o = f \left(\dfrac{1 \pm \dfrac{v_0}{v}}{1 \mp \dfrac{v_s}{v}} \right)$, where v_0 is velocity of observer, v_s is the velocity of the source, the frequency of the sound emitted by the source is f, and v is the velocity of the sound in the medium through which it is moving. In the numerator of the above expression, the + sign applies when the observer moves toward the source, and the minus sign is applicable when the observer moves away from the source. For the denominator, the positive sign applies when the source moves away from the observer, and the minus sign is used when the source moves toward the observer.

SOLVED PROBLEM

[Note: Unless otherwise stated, assume the speed of sound in air is 340 m/s]

❑❑ **If the speed of sound in air at 0 °C is 330 m/s, what is its speed when the temperature is 100 °C?**

$V = V_0 \sqrt{1 + \dfrac{t}{273}} = 330 m/s \sqrt{1 + \dfrac{100}{273}}$

$V \approx 386$ m/s at 100 °C.

❑❑ **Sound travels in water at about 1450 m/s. Find the Bulk modulus for water. (Assume the density of water is 1000 kg/m)**

For a fluid, $V = \sqrt{\dfrac{B}{P}} \Rightarrow V^2 = B/P$

Thus, $(1450 \text{ m/s})^2 (1000 \text{ kg/m}^3) = B$
$B = 2.1 \times 10^9 (m^2/s^2) \text{ kg/m}^3$
$B = 2.1 \times 10^9 \text{ N/m}^2$

❑❑ **What is the speed of sound in a steel rod if the density of the steel is 7800 kg/m³ and Young's modulus is 24 x 10¹⁰ N/m²?**

$V = \sqrt{\dfrac{Y}{P}} = \sqrt{\dfrac{24 \times 10^{10} N/m^2}{7800 kg/m^3}}$

$V = 5550$ m/s.

❑❑ **Suppose that the steel of the previous problem is used to make the rails of a railroad. If a worker drops a hammer on a rail, what would be the difference in time between the sound that travels through the air and the sound that travels through the steel for an observer 0.8 km away?**

For steel, $x = v_s t \Rightarrow t_s = \dfrac{x}{v_s} = \dfrac{800 m}{5550 m/s}$

$t_s = 0.144$ s

For air, $x = vt \Rightarrow t_{air} = \dfrac{x}{v_{air}} = \dfrac{800m}{340m/s}$

$$t_{air} = 2.35 \text{ s}$$

Thus, the difference $t_s - t_{air} = 2.21$ s.

❏❏ **Acoustic material in a room reduces the sound level by 25 dB. How much greater was the Intensity before the use of the acoustic material?**

$\beta = 10 \log \dfrac{I}{I_0} \Rightarrow 25 = 10 \log \dfrac{I}{I_0} \Rightarrow 316 = \dfrac{I}{I_0}$

so that before the reduction the Intensity was 316 times greater.

❏❏ **The threshold of pain in the ear because of the intensity of sound is about 120 dB. What is the intensity of this sound? (Assume that the threshold of hearing is 1×10^{-12} watt/m².)**

$$\beta = 10 \log \dfrac{I}{I_0}$$

$$120 = 10 \log \dfrac{I}{I_0}$$

$$1 \times 10^{12} = \dfrac{I}{I_0}$$

Thus, 120 dB corresponds to an intensity of 1 watt/m².

❏❏ **The intensity of the sound produced by a jet engine is 138 dB above the hearing threshold. Find the intensity of this sound if the hearing threshold intensity is 1×10^{-12} watt/m².**

$\beta = 10 \log \dfrac{I}{I_0} \Rightarrow \dfrac{138}{10} = \log \dfrac{I}{I_0}$

so that $(6.31 \times 10^{13})(1 \times 10^{-12}$ watt/m²$) = I = 63.1$ watt/m²

❏❏ **In electronics the signal-to-noise ratio (measured in dB) is important. Suppose that an electronic signal has an intensity of 1.5×10^{-4} watt/m², while the noise has an intensity level of 4.8×10^{-11} watt/m². By what number of decibels does the signal exceed the noise?**

$\beta = 10 \log \dfrac{I_s}{I_N} = 10 \log \dfrac{1.5 \times 10^{-4} watt/m^2}{4.8 \times 10^{-11} watt/m^2}$

$\beta = 64.9$ dB, i.e. the signal exceeds the noise by 64.9 decibels.

❏❏ **A certain hearing aid is advertised as increasing the sound level by 30 dB. By what factor does the intensity of the sound increase?**

$\beta = 10 \log \dfrac{I}{I_0} \Rightarrow 30 = 10 \log \dfrac{I}{I_0}$. Thus, $1000 = \dfrac{I}{I_0}$ so that the intensity of the sound increases by 1000 times.

❑❑ **At Band Day, a stationary observer is watching the band move toward her at a speed of 0.9 m/s. If a clarinet produces a sound that has a frequency of 784 Hz, what frequency does the observer hear?**

$f_0 = f \dfrac{\left(1 \pm \dfrac{v_0}{v}\right)}{\left(1 \mp \dfrac{v_s}{v}\right)}$. Here, $v_0 = 0$, and the v of sound in air is assumed to be 340 m/s. Since the sound is moving toward the spectator, $f_0 = 784\ Hz \left(\dfrac{1}{1 - \dfrac{0.9 m/s}{340 m/s}}\right) = 786 Hz$

The observer hears a frequency of 786 Hz.

❑❑ **For the previous problem, if the band does an about face and marches away from the observer, what frequency does the observer hear?**

Based on symmetry, the observer would hear a frequency of 782 Hz. From the general equation,

$f_0 = f \left(\dfrac{1}{1 + \dfrac{0.9 m/s}{340 m/s}}\right) = 782 Hz$

❑❑ **A bird is flying away from a stationary bird watcher at 10 m/s. If the bird produces a sound that has a frequency of 900 Hz, what is the frequency heard by the bird watcher?**

Since $f_0 = f \dfrac{\left(1 \pm \dfrac{v_0}{v}\right)}{\left(1 \mp \dfrac{v_s}{v}\right)}$, then since the bird is moving away from the bird watcher,

$f_0 = 900 Hz \left(\dfrac{1}{1 + \dfrac{10 m/s}{340 m/s}}\right)$ so that $f_0 = 874$ Hz. Hence the bird watcher hears a sound frequency of 874 Hz.

❑❑ **Two cars are moving toward each other with the same speed. If one car honks a horn with a frequency of 612 Hz, the observer in the second car hears a frequency of 635 Hz. What is the speed of each car?**

$f_0 = f \dfrac{\left(1 \pm \dfrac{v_0}{v}\right)}{\left(1 \mp \dfrac{v_s}{v}\right)}$

$635\ Hz = 612 \left(\dfrac{1 + \dfrac{v_0}{v}}{1 - \dfrac{v_s}{v}}\right)$

$$1.04 = \frac{1 + \frac{v_0}{340 m/s}}{1 - \frac{v_s}{340 m/s}} \text{ so that } 1.04 - \frac{1.04 v_s}{340 m/s} = 1 + \frac{v_0}{340 m/s}$$

$$.04 = \frac{2.04 v_0}{340 m/s}$$

Thus, $v_0 = v_s = 6.67$ m/s.

⬜ **Find the speed of sound in mercury. (B = 2.8 × 10¹⁰ N/m², and ρ = 13,600 kg/m³.**

$$V = \sqrt{\frac{Y}{\rho}} = \sqrt{\frac{2.8 \times 10^{10} N/m^2}{13,600 kg/m^3}} = 1435 m/s.$$

⬜ **For the previous problem, what would be the wavelength of a 2000 Hz sound wave in mercury?**

$$V = f\lambda \Rightarrow \lambda = \frac{V}{f} = \frac{1435 m/s}{2000 cy/s} = .717 m$$

⬜ **A toy train is moving on a circular track of radius 1.5 m and is making 2 rev/s. Its whistle sounds a continuous frequency of 1500 Hz. Find the maximum and minimum frequency heard by a stationary observer that is a large distance away.**

Here, the train is moving at 2 rev/s. But one rev = $2\pi r$ so that one rev = (6.28)(1.5 m) = 9.42 m.

Hence, $\frac{2 rev}{s} \left(\frac{9.42 m}{1 rev}\right) = 18.8$ m/s. The stationary observer will have the maximum frequency when the train is approaching, and the minimum frequency when the train is receding.

$$f_0 = f \left(\frac{1 \pm \frac{v_0}{v}}{1 \mp \frac{v_s}{v}}\right) \text{ and for the train approaching } f_0 = 2000 Hz \left(\frac{1}{1 - \frac{18.8 m/s}{340 m/s}}\right) \approx 2120 Hz$$

For the train receding $f_0 = 2000 Hz \left(\frac{1}{1 + \frac{18.8 m/s}{340 m/s}}\right) \approx 1900 Hz.$

⬜ **A sound source emits a sound with frequency 400 Hz. Find the observed frequency for each of the following situations: (a) observer at rest, sound approaching observer at 30 m/s, (b) observer moving toward an at rest source at 30 m/s, and (c) both source and observer moving toward each other at 15 m/s relative to the ground.**

(a) $f_0 = f \frac{\left(1 \pm \frac{v_0}{v}\right)}{\left(1 \mp \frac{v_s}{v}\right)}$

Here the observer is at rest, and the source is moving so

$$f_0 = 400 \text{ Hz} \left(\frac{1}{1 - \frac{30 m/s}{340 m/s}} \right) = 439 Hz$$

(b) Here, the observer is moving, and the source is at rest. Thus,

$$f_0 = 400 Hz \left(\frac{1 + \frac{30 m/s}{340 m/s}}{1} \right) = 435 Hz$$

(c) Both the source and observer are moving toward each other. Hence,

$$f_0 = 400 \text{ Hz} \left(\frac{1 + \frac{15 m/s}{340 m/s}}{1 - \frac{15 m/s}{340 m/s}} \right) = 437 Hz$$

☐☐ **A tuning fork emits a frequency of 441 Hz. A second tuning fork emits a frequency of 440 Hz. Find the beat frequency.**

The number of beats = $f_1 - f_2$. Hence, the beat frequency is 441 Hz - 440 Hz = 1 Hz.

☐☐ **A fire truck's siren has a frequency of 400 Hz, and the fire truck is approaching a concrete wall at 25 km/hr. (6.95 m/s) A stationary observer hears both the direct and reflected waves. Find the beat frequency.**

Case I: Observer in front of truck. Here, the observer will hear the sound from the truck and the sound reflected from the wall.

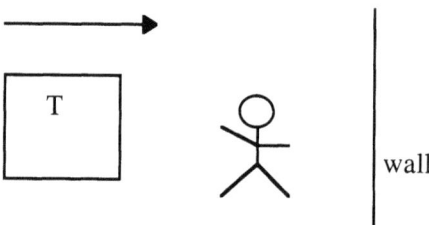

$$f_0 = f \left(\frac{1 \pm \frac{v_0}{v}}{1 \mp \frac{v_s}{v}} \right) = 400 Hz \left(\frac{1}{1 - \frac{6.95 m/s}{340 m/s}} \right) = 408 Hz$$

f_0 = 408 Hz for the direct wave, and 408 Hz for the reflected wave.
Thus the beat frequency = $f_1 - f_2$ = 408 Hz - 408 Hz = 0.

Case II: Observer behind the truck.

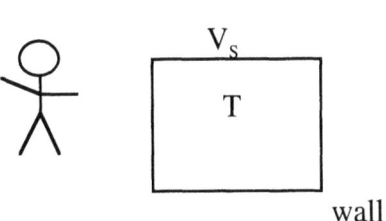

wall

$$f_0 = f\left(\frac{1}{1+\frac{V_s}{V}}\right) = 400 Hz \left(\frac{1}{1+\frac{6.95 m/s}{340 m/s}}\right) = 392 Hz$$

The frequency of the wave reflected from the wall is 408 Hz.
Hence, the beat frequency is $f_1 - f_2 = 408$ Hz $- 392$ Hz $= 16$ Hz.

□□ **An approaching source of sound emits a sound of 500 Hz. A stationary observer hears this sound at a frequency of 530 Hz. (a) Find the speed of the source, and (b) what frequency does the observer hear if the source recedes at the same speed?**

(a) $f_0 = 530$ Hz so that $f_0 = f\left(\frac{1\pm\frac{v_0}{v}}{1\mp\frac{v_s}{v}}\right) = 500 Hz \left(\frac{1}{1-\frac{v_s}{340 m/s}}\right)$, and thus,

$\left(1 - \frac{v_s}{340 m/s}\right) = \frac{500 Hz}{530 Hz} \Rightarrow v_s = 19.2 m/s$

(b) $f_0 = 500$ Hz $\left(\frac{1}{1+\frac{19.2 m/s}{340 m/s}}\right) = 473 Hz$

ELECTRICITY AND MAGNETISM
CHARGES AT REST

THE FUNDAMENTAL IDEAS ARE:

- Charge is a property of matter that causes it to both produce and experience electrical effects.

- There are two kinds of charge, positive and negative.

- Experiment shows that like charges repel, unlike charges attract.

- The units of charge are the coulomb and the atomic unit of charge, i.e., the charge of the proton and/or the charge of the electron. Note that 6.25×10^{18} atomic units of charge ≡ 1 C or the atomic unit of charge is $\pm 1.6 \times 10^{-19}$ C.

- A conductor is a material through which charge moves easily, an insulator is a material through which charge does not move easily. A semi-conductor is a material that when pure, behaves as an insulator, but with the addition of impurities will conduct.

- On an atomic level, a conductor is a material that has lots of free electrons, i.e., electrons that are loosely bound to the atom. These free electrons move easily throughout the metal. Insulators, however, have electrons that are tightly bound to the atom, and thus they do not move easily.

- The conservation of charge says that for an isolated system the total charge remains constant.

- Coulomb's Law is used to find the magnitude of the electric force that acts on one charge because it is in the presence of one or more charges, and is given by $F = \frac{kQ_1Q_2}{r_{12}^2}$, where Q_1 and Q_2 are the charges in Coulombs, r_{12} is the distance between the charges and k is the constant that is given by $k = 9 \times 10^9$ N-m²/C². k is also equal to $\frac{1}{4\pi\varepsilon_0}$, where ε_0 is 8.85×10^{-12} C²/N-m².

- Electric field is defined as any region where electric forces are observed, and has symbol $\vec{E}$.

- Electric field is a vector whose direction at any point is the same as the direction of the force that acts on a small positive test charge at the point in question.

- The mathematical definition of $\vec{E}_p$ is $\vec{E}_p = \frac{\vec{F}_p}{q}$, i.e., the force that acts on a small test charge at point p divided by the charge.

- Two results follow from this definition:
    $F_E = QE$ (the magnitude of the electric force is the magnitude of the charge times the magnitude of the electric field).

    At a point p, which is a distance r from a point charge Q, the magnitude of the electric field is given by $E_p = \frac{kQ}{r^2}$.

- Every charge creates an electric field; if there are several charges the resultant electric field in the vicinity of the charges is the vector sum of the electric fields produced by each charge.

- A line of force (or electric field line) is a line that is everywhere drawn in the direction of the electric field.

SOLVED PROBLEMS

☐☐ **An electroscope consists of a rod with two gold (or silver) "leaves" on one end. When a negatively charged rod is brought close to the electroscope, the leaves diverge (or separate). Explain why this occurs. Note that the leaves are very thin and thus are flexible.**

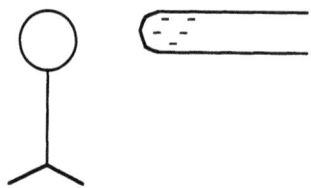

Since a metal is a conductor, there are lots of free electrons. The negatively charged rod causes these electrons to flow to the leaves, which thus become negatively charged, hence they repel each other.

☐☐ **How many electrons would have to be added to a nickel to give it a charge of -1μC?**

$Q_{total} = nq$, thus, $n = \dfrac{Q_{Total}}{q} = \dfrac{-1 \times 10^{-6} C}{-1.6 \times 10^{-19} C/e^{-}} = 6.25 \times 10^{12} e^{-}$

☐☐ **For the figure, find the force that acts on Q_1 because it's in the vicinity of Q_2**

$Q_1 = 6\mu C$

●——— 0.5 m ———●

$Q_2 = -4\mu C$

$F = \dfrac{kQ_1Q_2}{r^2} = \dfrac{(9 \times 10^9 N-m^2/C^2)(6 \times 10^{-6} C)(4 \times 10^{-6} C)}{(.5m)^2}$

$F = 0.864$ N. Thus, $\vec{F} = 0.864 N$ to the right. Note that Coulomb's Law gives us the magnitude of the force, and that is why the - for the -4μC was dropped.

☐☐ **An electron experiences a force of magnitude 6 x 10⁻¹⁶ N at a point. Find the magnitude of the electric field at the point.**

Since $E_p = \dfrac{F_p}{q} \Rightarrow E_p = \dfrac{6 \times 10^{-16} N}{1.6 \times 10^{-19} C} = 3750 N/C$. Note that this only gives the magnitude of the electric field, and this is why the − on the −1.6 × 10⁻¹⁹ C charge of an electron was dropped.

❏❏ **Find the electric field at point p in the figure, and the force that acts on a 1.6 x 10⁻¹⁴C charge at point p.**

$$Q = -12\mu C \bullet\text{─────────}\bullet p$$

$$E_p = \frac{kQ}{r^2} = \frac{(9x10^9 N-m^2)(12x10^{-6}C)}{(0.8m)^2} \approx 169,000 N/C$$

Thus, $\vec{E}_p \cong 169,000 N/C$ toward the negative charge. Again note that the expression used for the magnitude of the electric field gives the magnitude only. The direction of $\vec{E}$ is gotten by placing a small + test charge at the point in question, and finding the direction of the force that acts on the test charge. The direction of $\vec{E}$ is the same as the direction of the force that acts on the test charge.

$F_E = QE = (1.6 \times 10^{-14}$ C$)(169,000$ N/C$) = 2.7 \times 10^{-14}$ N. The direction on the force is to the left, or toward the -12µC charge.

❏❏ **Explain why for charges at rest all of the charge on a charged conductor lies on the surface of the conductor.**

The charges repel each other (like charges repel) as far away from each other as they can get. Hence, the charges lie on the surface of a conductor.

❏❏ **Consider the figure. Find the magnitude and direction of the electric field at point p.**

$$Q_1 = -6\mu C \qquad p$$
$$\bullet\text{─────}\bullet\text{────────}\bullet$$
$$\qquad 0.1 m \qquad 0.2 m$$
$$\qquad\qquad\qquad Q_2 = 4\mu C$$

$\vec{E}_p = \vec{E}_{pQ_1} + \vec{E}_{pQ_2}$. By placing a small + test charge at point p, the direction of both electric fields will be to the left.

$$E_{pQ_1} = \frac{kQ_1}{r_{pQ_1}^2} = \frac{(9x10^9 N-m^2/C^2)(6x10^{-6}C)}{(.1m)^2} = 5.4x10^6 N/C$$

$$E_{pQ_2} = \frac{kQ_2}{r_{pQ_2}^2} = \frac{(9x10^9 N-m^2/C^2)(4x10^{-6}C)}{(.2m)^2} = 0.9x10^6 N/C$$

Thus, $\vec{E}_p = 6.3x10^6 N/C$ to the left.

❏❏ **Explain why for two positive (or negative) charges the only place the $\vec{E}$ can be 0 on the line connecting the two charges is between the two charges.**

Consider the figure:

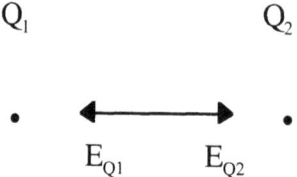

Only in the region between the two charges will the electric fields point in the opposite direction. Thus, this is the only place where the vector sum can be 0.

❏❏ **Consider the figure. Find the $\vec{E}$ at the origin.**

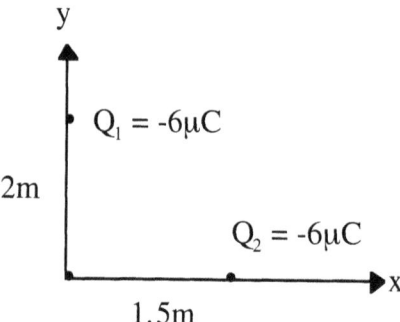

Here, each charge produces an electric field. Thus,

$$E_{0Q_1} = \frac{kQ_1}{r^2} = \frac{(9x10^9)(6x10^{-6}C)}{(2m)^2} = 13,500 N/C$$

$$E_{0Q_2} = \frac{kQ_2}{r^2} = \frac{(9x10^9)(6x10^{-6}C)}{(1.5m)^2} = 24,000 N/C$$

Since $\vec{E}$ is a vector, one needs to find both magnitude and direction.

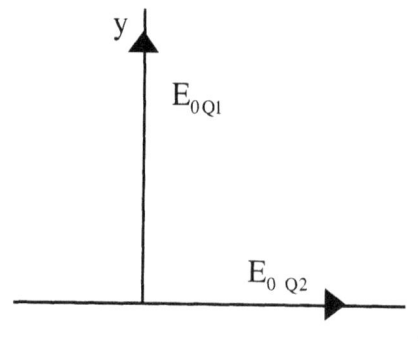

Using the vector addition table:

Vector	x	y
E_{0Q1}	0	13,500 N/C
E_{0Q2}	24,000 N/C	0
E_0	24,000 N/C	13,500 N/C

Thus, the table tells us the component of the resultant.

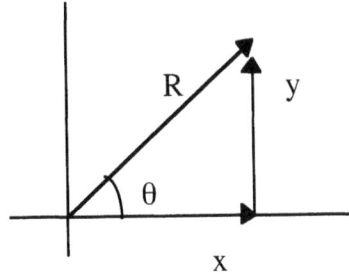

so that $E_0 = \sqrt{x^2 + y^2} = \sqrt{(24,000 N/C)^2 + (13,500 N/C)^2} = 27,500 N/c$

$\tan \theta = \dfrac{y}{x} = \dfrac{13,500 N/C}{24,000 N/C}$

$\theta = 29.4°$

so that finally $\vec{E}_0 \cong 27,500 N/C$ @ $29.4°$
above + x axis.

☐☐ (a) What is the magnitude of the force that would act on an electron at the origin of the above problem, and (b) what acceleration acts on the electron when it is at the origin of the above problem? (The mass of an electron is 9.1 x 10⁻³¹ kg.)

(a) $F = QE \Rightarrow F = (1.6 \times 10^{-19} C)(27,500 N/C) = 4.4 \times 10^{-15} N$

(b) $F = ma \Rightarrow QE = ma$ so that $a = \dfrac{QE}{m} = \dfrac{4.4 \times 10^{-15} N}{9.1 \times 10^{-31} kg}$

$a = 4.84 \times 10^{15} m/s^2$

☐☐ Find the force that acts on Q_1 because of Q_2 and Q_3 in the figure.

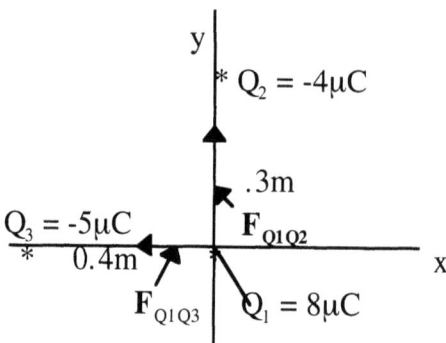

$\vec{F}_{Q_1} = \vec{F}_{Q_1Q_2} + \vec{F}_{Q_1Q_2}$

$F_{Q_1Q_2} = \dfrac{kQ_1Q_2}{r^2} = \dfrac{(9x10^9 N-m^2/C^2)(8x10^{-6}C)(4x10^{-6}C)}{(0.3m)^2} = 3.2N$

$F_{Q_1Q_3} = \dfrac{kQ_1Q_3}{r^2} = \dfrac{(9x10^9 N-m^2/C^2)(8x10^{-6}C)(5x10^{-6}C)}{(0.4m)^2} = 2.25N$

Thus,

Vector	x	y
$F_{Q_1Q_2}$	0	3.2N
$F_{Q_1Q_3}$	-2.25N	0
R	-2.25N	3.2N

$R = \sqrt{x^2 + y^2} = \sqrt{(-2.25N)^2 + (3.2N^2)} \Rightarrow R = 3.91N$

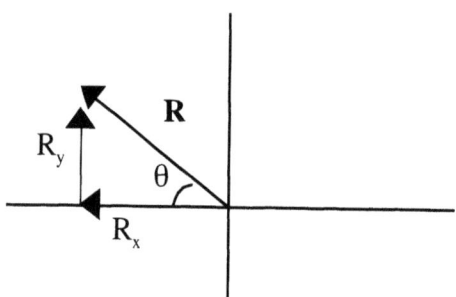

$\tan\theta = \dfrac{R_y}{R_x} = \dfrac{3.2N}{2.25N} = 1.42 \Rightarrow \theta = 54.9°$

Thus, $\vec{F}_{Q_1}$, because of Q_2 and Q_3, = 3.91N @ 54.9° above - x axis.

☐☐ Consider the figure:

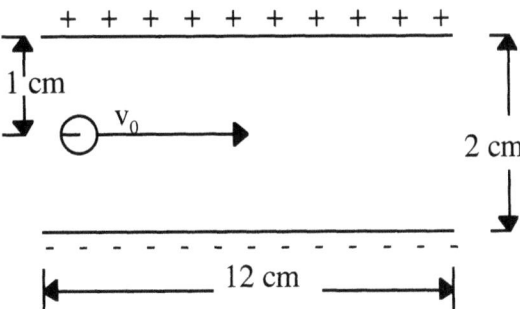

An electron moving with horizontal velocity of 2 x 10⁶ m/s enters the region between two parallel metal plates. The plates are separated by 2 cm, and are 12 cm long. For what value of $\vec{E}$ will the electron just miss the upper plate when it leaves the region between the plates? (The mass of an electron is 9.1 x 10⁻³¹ kg)

This is a two-dimensional motion problem with acceleration in y caused by the electric force that acts on the electron. To find the time the electron is between the plates, $x = v_{av}t \Rightarrow .12\ m = (2 \times 10^6\ m/s)t$
Thus, $t = 6 \times 10^{-8}$ s.

For the y component of the motion, $F = QE = ma$. Thus $a_y = \dfrac{QE}{m}$.

When the electron leaves the region between the plates, it just misses the upper plate. Thus, $y = \dfrac{1}{2}a_y t^2 + (v_0)_y t$. Since $(v_0)_y = 0$, $\dfrac{2y}{t^2} = a = \dfrac{QE}{m}$, so that

$\dfrac{2ym}{t^2 Q} = E = \dfrac{2(.01m)(9.1 \times 10^{-31} kg)}{(6 \times 10^{-8} s)^2 (1.6 \times 10^{-19} C)}$

E = 31.6 N/C. The direction of E is toward the bottom of the page.

☐☐ For the previous problem, suppose that a proton is released from rest at the upper plate. How long would it take the proton to strike the lower plate? (the mass of a proton is 1.67 x 10⁻²⁷ kg)

$F = QE = ma \Rightarrow a = Q\dfrac{E}{m} = \dfrac{(1.6 \times 10^{-19} C)(31.6 N/C)}{(1.67 \times 10^{-27} kg)}$

$a = 3.03 \times 10^9\ m/s^2$ down.

Thus, $y = \dfrac{1}{2}at^2 + v_0 t \Rightarrow -0.2m = \dfrac{1}{2}(-3.03 \times 10^9\ m/s^2)t^2$

$t' = 1.32 \times 10^{-10} s \Rightarrow t = 1.15 \times 10^{-5} s$

❏❏ Two charges, $Q_1 = 8\,\mu C$ and $Q_2 = 6\,\mu C$ are 0.9 m apart. Where on the line joining the two charges is the resultant field 0?

```
Q₁              P              Q₂
*───────────────*───────────────*
       x           (.9 m - x)
```

Since the only place the two electric fields oppose each other is between the two charges, let Q_1 be at the origin where $x = 0$.
$E_P = 0 \Rightarrow E_{PQ_1} = E_{PQ_2}$

Thus, $\dfrac{kQ_1}{x^2} = \dfrac{kQ_2}{((.9m-x)^2} \Rightarrow \dfrac{Q_2}{Q_1} = \dfrac{x^2}{(.9m-x)^2} = \dfrac{6\mu C}{8\mu C}$

Thus, $\dfrac{x}{.9m-x} = \pm\sqrt{.75} = \pm.866$

Case I: $\dfrac{x}{.9m-x} = .866 \Rightarrow x = .779m - .866x$, so that $1.866x = .779m$ and $x = .417m$

Case II: $\dfrac{x}{.9m-x} = -.866 \Rightarrow x = -.779m + .866x$, so that $.134x = -.779$ and $x = -5.81m$.

Since, case II is physically impossible, the resultant electric field will be 0 at 0.417 m from Q_1 between Q_1 and Q_2.

❏❏ Two charges, $Q_1 = Q_2 = Q$ are placed on the y axis at $(0,a)$ and $(0,-a)$. (a) Show that the electric field at any point P on the x-axis is directed along the x-axis, (b) find the expression for the magnitude of the electric field at any point P on the x-axis, (c) what is the electric field magnitude when $x = 0$, and (d) as x gets very large what is the expression for the magnitude of E_P?

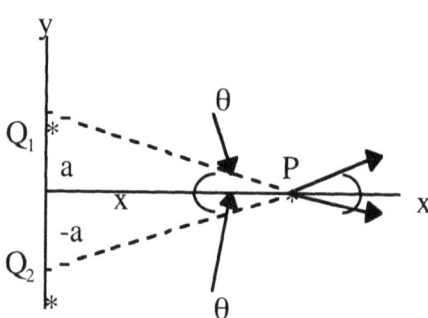

(a) Because of the symmetry, the y components of the electric field at P oppose each other and sum to 0. Thus, only the x components survive, and hence, $\vec{E}_P$ is directed along the positive x-axis.

(b) $E_{PQ_1} = \dfrac{kQ}{x^2 + a^2}$ and $E_{PQ_2} = \dfrac{kQ}{x^2 + a^2}$.

Thus, $\left(E_{PQ_1}\right)_x = \left(E_{PQ_2}\right)_x = \left(\dfrac{kQ}{x^2+a^2}\right)\cos\theta = \left(\dfrac{kQ}{x^2+a^2}\right)\dfrac{x}{\sqrt{a^2+x^2}}$

so that $E_P = \dfrac{2kQx}{\left(x^2+a^2\right)^{\frac{3}{2}}}$

(c) If $x = 0$, $E_P = 0$, as can be seen by looking at the direction of each field.

(d) If x tends to a large number, then $E_P \approx \dfrac{2kQx}{x^3} \approx \dfrac{2kQ}{x^2}$. From a large distance the two charges appear to be a single charge 2Q.

☐☐ A Cottrell electrostatic precipitator consists of a charged wire (or rod) that is placed down a smokestack. The result is that emissions are greatly reduced. What is the idea behind this device?

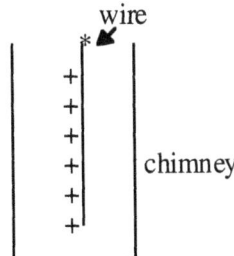

The idea is that $F_E = QE$. If the pollutants or particulate matter have a charge, the wire, which is charged, creates an electric field. Thus, there is an electric force that causes the charged particles to be either repelled away from the wire or toward the wire. For cleaning purposes, it is usually easier to have the charged particles go toward the wire, rather than have them be repelled toward the walls of the chimney. Note that if the charge on the pollutants or particulate matter is 0, the electric force is 0 and hence the device will have no effect on smokestack emissions.

☐☐ A proton with velocity of 3×10^7 m/s enters a uniform electric field of magnitude 50,000 N/C as if the figure. Find how far the proton moves before coming to rest by (a) using Newton's Laws and equation of motion approach, and (b) by using an energy approach. (The mass of the proton is 1.67×10^{-27} kg)

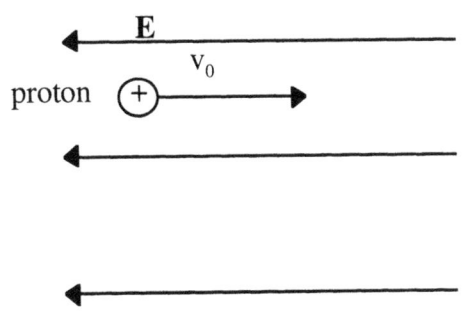

(a) Newton's Laws and equation of motion approach, $F = ma = QE$. Thus,

$$a = \frac{QE}{m} = \frac{(1.6 \times 10^{-19} C)(50,000 N/C)}{1.67 \times 10^{-27} kg}$$

$a = 4.79 \times 10^{12}$ m/s² to the left.

Since, $v = v_0 + at \Rightarrow \frac{v - v_0}{a} = t = \frac{0 - 3 \times 10^7 m/s}{-4.79 \times 10^{12} m/s^2}$, so that $t = 6.26 \times 10^{-6}$ s.

Since, $x = v_{av} t \Rightarrow x = \left(\frac{0 + 3 \times 10^7 m/s}{2}\right)(6.26 \times 10^{-6} s)$

$x = 93.9$ m.

(b) Energy approach: Change in energy = work done
$E - E_0 = Fr\cos\theta$
$0 - (1/2)mv_0^2 = QEr\cos 180°$

so that $\frac{-\frac{1}{2}mv_0^2}{QE\cos\theta} = r = \frac{-.5(1.67 \times 10^{-27} kg)(3 \times 10^7 m/s)^2}{(1.6 \times 10^{-19} C)(50,000 N/C)(-1)}$

$r = 93.9$ m.

Either method says that the proton moves 93.0 m before coming to rest.

☐☐ Two pith balls (mass of each is 2 g) are given an equal charge Q as in the figure. Find the charge on either ball. The threads that support each ball are 20 cm long and each thread makes an angle of 20° with respect to the vertical.

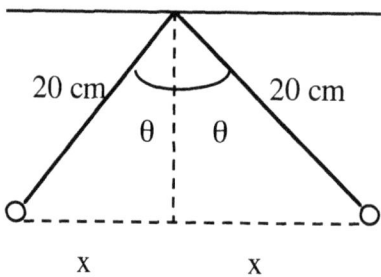

Since the pith balls are at rest, the conditions of equilibrium apply.
Thus, $\Sigma F_{up} = \Sigma F_{down}$ and $\Sigma F_{left} = \Sigma F_{right}$, so that $T_V = mg$ and $T_H = F_E$.

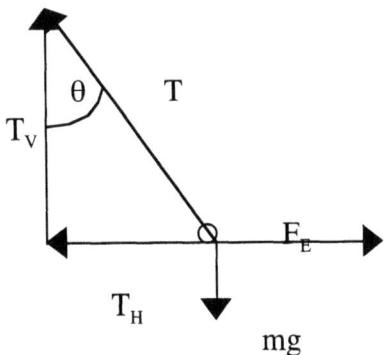

$T_V = T\cos\theta$ and $T_H = T\sin\theta$
$x = (.2\ m)\tan\theta = .0727\ m$
$r = .146\ m$

Rewriting, (1) $\dfrac{kQ^2}{r^2} = T_H = T\sin\theta$

(2) $mg = T_V = T\cos\theta$

Dividing equation (1) by equation (2), $\dfrac{\frac{kQ^2}{r^2}}{mg} = \tan\theta$

or, $Q^2 = \dfrac{r^2 mg \tan\theta}{k} = \dfrac{(0.146m)^2(.002)(9.8)N\tan 20°}{9 \times 10^9\ N-m^2/C^2}$

$Q^2 = 1.68 \times 10^{-14}\ C$
$Q = 1.3 \times 10^{-7}\ C$

POTENTIAL

THE IDEAS HERE ARE THE FOLLOWING:

- In the vicinity of a charge Q, the potential V is defined to be $V_P = \dfrac{U_P}{q}$, where the 0 of U is at ∞, and q is a charge that has been moved from ∞ to point P.

- $V_P = \dfrac{W_{\infty \to P}}{q} = \dfrac{kQ}{r}$

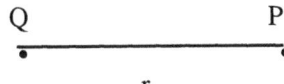

  where V has units of Joules/Coulomb or volt. Note that positive charge creates positive potential while negative charge creates negative potential.

- In two dimensions, a line that has the same potential is called an equipotential line; in three dimensions a surface that has the same potential is called an equipotential surface.

- Every charge creates a potential (as well as an electric field). If there are several charges in the vicinity of point P, then the potential at point P, V_P, is found by $V_P = V_{PQ_1} + V_{PQ_2} + Q_{PQ_3} + ...$

- The potential difference between two point A and B in the figure may be calculated in several different ways:
$V_{AB} = V_A - V_B = \dfrac{kQ}{r_A} - \dfrac{kQ}{r_B} = \dfrac{W_{B \to A}}{Q}$

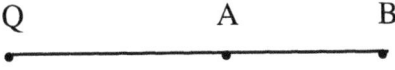

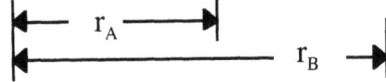

- Potential difference is measured with a voltmeter, and the schematic representation of a voltmeter is as noted:

- The electric field between two parallel metal plates, one with + Q and the other with − Q, is in magnitude given by $E = \dfrac{V}{d}$, where V is the potential difference between the plates, and d is the distance between the plates.

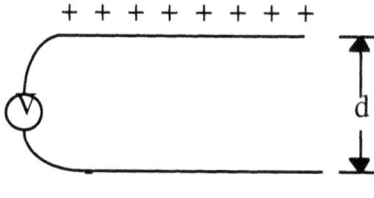

- Since $V = \dfrac{W}{Q} \Rightarrow W = VQ =$ work done on a charge Q that moves through a potential difference.

- Work done equals change in energy, thus, VQ = work done = ΔEnergy of the charge.

- Robert Millikan (and his students), from 1909 - 1913, used the ideas of the electric force acting on a charged particle between two parallel metal plates to find the charge of an electron.

- An oscilloscope is used to give a visual representation of an electronic signal by passing a beam of electrons through two sets of parallel metal plates.

- If an electron moves through a potential difference of one volt, then W = ΔEnergy = QV = 1ev (eV is electron volts). 1 eV = 1.6 x 10^{-19} J. A 10 eV electron has an energy of 10 eV or 1.6 x 10^{-18}J.

- The potential at the surface of a conducting sphere is the same as if all of the charge were at the center of the sphere, and the potential within the sphere is the same as the potential at the surface of the sphere.

SOLVED PROBLEMS

❏❏ Point P is 0.6 m from an 8µC charge as in the figure. Find (a) the potential at point P, and (b) the amount of work needed to bring a proton from ∞ to point P.

(a) $V_P = \dfrac{kQ}{r} = \dfrac{(9 \times 10^9 N - m^2/C^2)(8 \times 10^{-6} C)}{.6m} = 120,000V$

(b) $V_P = \dfrac{W_{\infty \to P}}{q} \Rightarrow W = (V_P)q = (120,000 J/C)(1.6 \times 10^{-19} C)$

$W = 1.92 \times 10^{-14}J$

☐☐ Point P is located between two charges Q_1 and Q_2 as in the figure. Find (a) the potential at point P, and (b) the work needed to bring a charge of 7 nC from ∞ to point P.

$$Q_1 = -6\mu C \qquad P \qquad Q_2 = 7\mu C$$
$$\bullet\underset{0.3\ m}{\rule{3cm}{0.4pt}}\bullet\underset{0.2\ m}{\rule{2cm}{0.4pt}}\bullet$$

(a) $V_P = V_{PQ_1} + V_{PQ_2} = \dfrac{kQ_1}{r_1} + \dfrac{kQ_2}{r_2} = \dfrac{(9x10^9 N-m^2/C^2)(-6x10^{-6}C)}{0.3m} + \dfrac{(9x10^9 N-m^2/C^2)(7x10^{-6}C)}{0.2m}$

$V_P = 135,000$ V

(b) $V_P = \dfrac{W_{\infty \to P}}{g} \Rightarrow W = V_P g = (135,000V)(7x10^{-9}C)$

$W = 9.45 \times 10^{-4}$ J

☐☐ Consider the figure. Find the electric field in the region between the two metal plates.

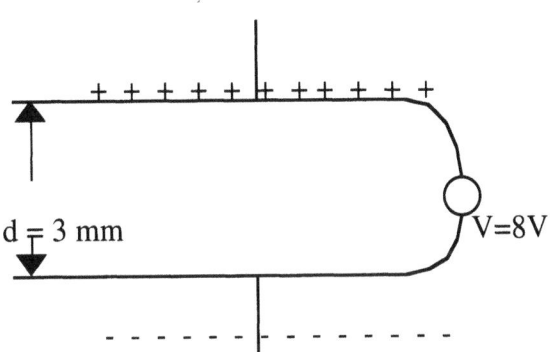

$E = \dfrac{V}{d} = \dfrac{8V}{3x10^{-3}m} \cong 2670V/m$

☐☐ For the previous problem, a proton (mass = 1.67×10^{-27} kg) is released from rest at the upper plate. Find the speed with which it hits the lower plate by (a) using a Newton's Laws approach, and (b) using an energy approach. Assume that gravitational effects are negligible.

(a) Newton's Laws approach: since $F = QE \Rightarrow ma = QE$ so that $a = (QE)/m$,

$a = \dfrac{(1.6x10^{-19}C)2670V/m}{1.67x10^{-27}kg} = 2.56x10^{11} m/s^2$

Using motion tool (4), $v^2 = v_0^2 + 2ax = 0 + 2(2.56 \times 10^{11}$ m/s$^2)(3 \times 10^{-3}$ m)
$v^2 = 1.54 \times 10^9$ m^2/s$^2 \Rightarrow v = \pm 3.92 \times 10^4$ m/s

The speed of the proton as it impacts the lower plate is 3.92×10^4 m/s

(b) Energy approach:

Here $V = \dfrac{work}{Q} \Rightarrow work = VQ = \Delta Energy$

Thus, $(1/2)mv^2 - 0 = VQ$, so that $v = \sqrt{\dfrac{2VQ}{m}} = \sqrt{\dfrac{2(8V)(1.6x10^{-19}C)}{1.67x10^{-27}kg}}$

$v = \pm\ 3.92 \times 10^4$ m/s

So the speed of the proton as it strikes the lower plate is 3.92×10^4 m/s.
Note that one could also make use of the fact that when the proton impacts the lower plate it will acquire an energy of 8 eV; thus $(1/2)mv^2 = 8eV = 1.28 \times 10^{-18}$ J, so that

$v = \sqrt{\dfrac{2(1.28x10^{-18}J)}{1.67x10^{-27}kg}} = 3.92x10^4 m/s$ as before.

❏❏ A conducting sphere of radius 35 cm has -15 µC of charge on its surface. Find the potential (a) at the surface of the sphere and at point P within the sphere.

(a) $V = \dfrac{kQ}{R} = \dfrac{(9x10^9 N-m^2/C^2)(-15x10^{-6}C)}{35m}$

V = -3.86 x 10⁵ V at the surface of the sphere.

(b) The potential within the sphere is the same as the potential at the surface
V_P = -3.86 x 10⁵ V.

❏❏ For the previous problem, find (a) the potential difference between point A and point B (note the figure), and (b) the work done in moving an alpha particle (charge 3.2 x 10⁻¹⁹ C) from B to A.

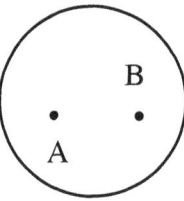

(a) V_{AB} = 0 (both points are at the same potential)

(b) Since $V_{AB} = \dfrac{W_{B \to A}}{Q} \Rightarrow W_{B \to A} = 0$

❏❏ Consider the figure.. Find V_{AB}

-16 µC   A   B

|← .4 m →|← .2m →|

$$V_{AB} = V_A = V_B = \frac{kQ}{r_A} - \frac{kQ}{r_B}$$

$$V_{AB} = \frac{(9x10^9 N - m^2/C^2)(-16x10^{-6}C)}{0.4m} - \frac{(9x10^9 N - m^2/C^2)(-16x10^{-6}C)}{0.6m}$$

$V_{AB} = -1.2 \times 10^5$ V

☐☐ For the previous problem, (a) what is the work done to move an electron from B → A, and (b) what would a voltmeter that is connected between points A and B read?

(a) $V = \frac{W}{Q} \Rightarrow W_{B \to A} = (V_{AB})Q = (-1.2x10^5 V)(-1.6x10^{-19}C)$

$W_{B \to A} = 1.92x10^{-14} J$

(b) Since a voltmeter measures potential difference, the voltmeter would read 1.2 x 10⁵ V (it could read – 1.2 x 10⁵ V if it were a digital meter).

☐☐ The purpose of a pair of accelerating plates is to give a charged particle acceleration and hence a change in velocity. For the figure, find (a) the acceleration that acts on the electron, (b) the work done on the electron as it reaches the second plate, (c) the speed of the electron when it reaches the second plate, and (d) the magnitude of the electric field between the parallel metal plates. Assume that the electron starts from rest. (The mass of an electron is 9.1 x 10⁻³¹ kg.)

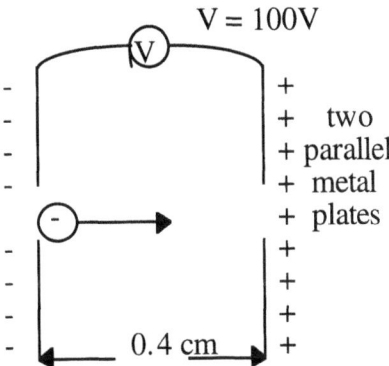

(a) Since the work done on the electron equals the change in energy, work = VQ = (100V)(1.6 x 10⁻¹⁹C). The work done in magnitude is 1.6 x 10⁻¹⁷ J. Hence, 1.6 x 10⁻¹⁷ J = k - k₀ or 1.6 x 10⁻¹⁷ J =
(1/2)mv² ¿ 0 ⇒ $\sqrt{\frac{2(1.6x10^{-17}J)}{9.1x10^{-31}kg}} = v = 5.93x10^6 m/s$

Using motion tool 1, x = (v_av)t ⇒ $\frac{4x10^{-3}m}{2.97x10^6 m/s} = t$

t = 1.35 x 10⁻⁹s
Thus, from motion tool 3, v = v₀ + at ⇒ 5.93 x 10⁶ m/s = 0 + a(1.35 x 10⁻⁹s) and
a = 4.4 x 10¹⁵ m/s²

(b) From above, work done = 1.6 x 10⁻¹⁷ J.

(c) From above, the speed of the electron when it reaches the second plate is 5.93 x 10⁶ m/s.

(d) First way: $E = \dfrac{V}{d} = \dfrac{100V}{4 \times 10^{-3} m} = 25{,}000 N/C$

Second way: $F = QE = ma$, and thus, $E = \dfrac{ma}{Q} = \dfrac{(9.1 \times 10^{-31} kg)(4.4 \times 10^{15} m/s^2)}{-1.6 \times 10^{-19} C}$

$E = -25{,}000$ N/C

Thus, $\vec{E} = 25{,}000$ N/C directed toward the negatively charged plate.

☐☐ Show that the electric force on a charged particle that is between two parallel plates is proportional to the potential difference between the plates.

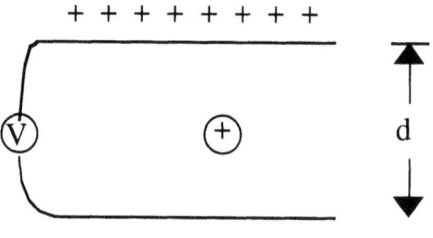

Since $F = QE$ and $E = \dfrac{V}{d}$ for the electric field in the region between the plates, $F = Q\dfrac{V}{d}$.
If the distance between the plates and the charge of the particle are constant, $F \sim V$. Hence, the bigger the potential difference, the bigger the electric force.

☐☐ Two 8 μC charges are at opposite ends of the hypotenuse of a 3 m, 4 m, 5 m right triangle. Find (a) the potential at the third vertex (corner) of the triangle and (b) find the work done in moving an alpha particle (charge +2e) from infinity to the third vertex.

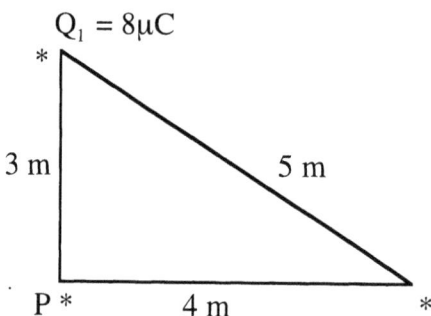

(a)
$V_P = V_{PQ_1} + V_{PQ_2}$

$V_P = \dfrac{kQ_1}{r_1} + \dfrac{kQ_2}{r_2}$

$V_P = \dfrac{(9 \times 10^9 N\text{-}m^2/C^2)(8 \times 10^{-6} C)}{3m} + \dfrac{(9 \times 10^9 N\text{-}m^2/C^2)(8 \times 10^{-6} C)}{4m}$

$V_P = 42{,}000$ V at the third vertex

(b) $V_P = \dfrac{W_{\infty \to P}}{Q} \Rightarrow W_{\infty \to P} = (V_P)Q = (42,000 J/C)(2)(1.6 \times 10^{-19} C)$

$W_{\infty \to P} = 1.34 \times 10^{-14} J$ is the work needed to bring the alpha particle from infinity to the third vertex of the right triangle.

☐☐ **An equipotential surface of potential 30 V surrounds a point charge of 3×10^{-9} C. Find the radius of the equipotential surface.**

$v = \dfrac{kQ}{r}$, and thus $r = \dfrac{kQ}{V} = \dfrac{(9 \times 10^9 N - m^2/C^2)(3 \times 10^{-9} C)}{30 V}$

$r = 0.9$ m

☐☐ **A conducting droplet with a radius of 0.5 mm has a charge of 500 protons. The droplet combines with an identical droplet, also having a charge of 500 protons. Find the potential at the surface of the new larger droplet.**

At the surface of a conducting sphere, $V = \dfrac{kQ}{R}$, where R is the radius of the sphere. Thus, one can find the potential at the surface of the new droplet if the radius of the new droplet is known. Since the volume of a sphere is given by $(4/3)\pi r^3$,
$V_{new} = V_1 + V_2 \Rightarrow (4/3)\pi r_N^3 = 2(4/3)\pi r_1^3$) or, $r_N = (2)^{1/3} r_1 = (2)^{1/3}(5 \times 10^{-4} m^3)$
$r_N = 6.3 \times 10^{-4}$ m, and the new charge is 1000 protons.

Thus, $V_{surface} = \dfrac{kQ}{r_N} = \left(\dfrac{9 \times 10^9 N - m^2/C^2}{6.3 \times 10^{-4} m}\right)(1000)(1.6 \times 10^{-19} C)$

$V_{surface} = 2.29 \times 10^{-3}$ V = 0.00229 V

CAPACITORS

THE FUNDAMENTAL IDEAS ARE THE FOLLOWING:

- A capacitor is a device that is used to store charge (or energy)

- The usual configuration for a capacitor is two conductors separated by an insulator, or dielectric.

- The mathematical definition of capacitors is $C = \frac{Q}{V}$, where C has units of Farads. Note that $1 F = \frac{1C}{V} = \frac{1C^2}{J}$.

- The capacitance of a parallel plate capacitor with air between the plates is given by
- $C = \frac{\varepsilon_0 A}{d}$, where ε_0 is a constant equal to $8.85 \times 10^{-12} C^2/N\text{-}m^2$. A is the area of one of the plates, and d is the distance between the two plates.

- The ratio of the capacitance of a capacitor with a dielectric between its plates to the capacitance without the material between the plates is called the dielectric constant K. Thus, $K = \frac{C}{C_0}$. The dielectric material increases the capacitance of the capacitor and aids in the manufacturing process by allowing a uniform spacing between the plates of a capacitor.

- The dielectric strength of any material is the maximum electric field that can be applied to a material before it loses insulation properties. For the material between the plates of a capacitor, dielectric strength = $E_{max} = \frac{V_{max}}{d}$ allows one to find the maximum potential difference that can be applied to the plates of a capacitor without dielectric breakdown. This is the reason that every capacitor is given two numbers: its capacitance value and the maximum voltage that can be applied to the plates of the capacitor without dielectric breakdown.

- The energy stored in a capacitor can be found by $U_C = (1/2)CV^2 = (1/2)QV = \frac{Q^2}{2C}$.

- The energy density of the electric filed is given by $U_E = (1/2)\varepsilon_0 E^2$, where $\varepsilon_0 = 8.85 \times 10^{-12} C^2/N\text{-}m^2$.

- For capacitors hooked in series, one can find the equivalent (or total) capacitance by
$\frac{1}{C_{eq}} = \frac{1}{C_1} + \frac{1}{C_2} + \frac{1}{C_3} + ...$

- For capacitors hooked in parallel, one can find the equivalent (or total) capacitance by
$C_{eq} = C_1 + C_2 + C_3 + ...$

- For capacitors hooked in series, the charge is the same for each capacitor.

- For capacitors hooked in parallel, the total charge is the sum of the charges on each capacitor.

SOLVED PROBLEMS

❏❏ **Assuming that the earth is a conducting sphere with a charge Q, find its capacitance. (The radius of the earth is ≈ 6400 km)**

$$C = \frac{Q}{V} = \frac{Q}{k\frac{Q}{R}} = \frac{R}{k}$$

$$C = \frac{6.4 \times 10^6 \, m}{9 \times 10^9 \, N-m^2/C^2} = 7.11 \times 10^{-4} \, C^2/J$$

C = 711 μF

❏❏ **A parallel plate capacitor has a capacitance of 1 F. If the plates are separated by 1 mm, how big is the area of one of the plates?**

$$C = \varepsilon_0 \frac{A}{d} \Rightarrow \frac{Cd}{\varepsilon_0} = A = \frac{(1F)(1 \times 10^{-3} m)}{8.85 \times 10^{-12} \, C^2/N-m^2}$$

A = 1.13 x 10^8 m^2 = 1.13 x 10^2 km^2
Note that a capacitor that has a capacitance of 1 F is a very large capacitor.

❏❏ **For the previous problem, assume that the dielectric strength of air is 3 x 10^6 V/m. Find the maximum potential difference that can be applied to the plates of the capacitor without dielectric breakdown.**

Dielectric strength = $E_{max} = \frac{V_{max}}{d}$.

Thus, $V_{max} = (E_{max})d$ = (3 x 10^6 V/m)(1 x 10^{-3} m)

V_{max} = 3000 V

❏❏ **Waxed paper (dielectric constant = 3.5) is inserted between the plates of an air capacitor that had value of 4 μF. Find the new value of the capacitance.**

$K = \frac{C}{C_0} \Rightarrow C = KC_0$ and thus, C = (3.5)(4 μF) = 14 μF

☐☐ Find the capacitance between points A and B of the figure. $C_1 = C_2 = C_3 = 10\ \mu F$

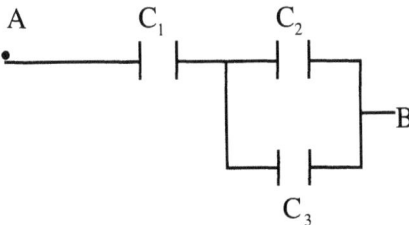

Since C_2 and C_3 are in parallel, $C_{EQ} = 20\ \mu F$. Thus, the circuit reduces to

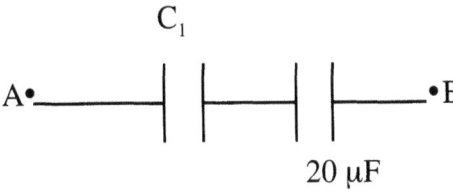

so that the 20 µF capacitor and C_1 are in series. Thus,

$$\frac{1}{C_{AB}} = \frac{1}{C_1} + \frac{1}{20\mu F} = \frac{1}{10\mu F} + \frac{1}{20\mu F}$$

$\frac{1}{C_{AB}} = \frac{3}{20\mu F}$, so that $C_{AB} = 6.67\ \mu F$

☐☐ For the previous problem, if a potential difference of 15 V is applied between points A and B, find the energy stored in the capacitor combination.

$U_C = (1/2)CV^2 = (1/2)(6.67 \times 10^{-6}\ F)(15V)^2$

$U_C = 7.5 \times 10^{-4}\ J = .75\ mJ$

☐☐ A 16 µF, 4000 V capacitor has waxed paper (dielectric constant 3.5, dielectric strength 40×10^6 V/m) between its plates. Find (a) the thickness of the waxed paper between the plates, (b) the energy stored in the capacitor and (c) the area of one of the plates if 3500 V is applied to the capacitor.

(a) Dielectric strength = $E_{max} = \frac{V_{max}}{d}$, so that 40×10^6 V/m = $\frac{4000V}{d}$

$d = 1 \times 10^{-4}\ m = 0.1\ mm$

(b) $U_C = (1/2)CV^2 = (1/2)(16 \times 10^{-6}\ F)(3500\ V)^2$
$U_C = 98\ J$

(c) $C = KC_0 = \dfrac{K\varepsilon_0 A}{d}$

16×10^{-6} F $= 3.5(8.85 \times 10^{-12}$ C^2/N-m^2) $\dfrac{A}{1 \times 10^{-4} m}$

A = 51.7 m^2

☐☐ A parallel plate capacitor has air between the plates (dielectric strength 3 x 10^6 V/m) and is rated at 50 µF, 500 V. Find (a) the maximum energy the capacitor can store, (b) the maximum charge that can be stored in the capacitor, and (c) the electric field between the plates when 500 V is applied to the capacitor, (d) the energy density between the plates when 500 V is applied to the capacitor and (e) the thickness of the air space between the plates.

(a) $U_C = (1/2)CV^2 = (1/2)(50 \times 10^{-6}$ F$)(500$ V$)^2$
$U_C = 6.25$ J

(b) $C = \dfrac{Q}{V} \Rightarrow Q = CV$

Thus, the maximum charge is stored in the capacitor when the maximum potential difference is applied to capacitor.
$Q_{max} = Cv_{max} = (50 \times 10^{-6}$ F$)(500$ V$) = 2.5 \times 10^{-2}$ C

(c) Dielectric strength $= E_{max} = \dfrac{V_{max}}{d} \Rightarrow d = \dfrac{500V}{3 \times 10^6 V/m}$

d = 1.667 x 10^{-4} m = .167 mm
Thus, $E = \dfrac{V}{d} = \dfrac{500V}{1.667 \times 10^{-4} m} \cong 3 \times 10^6 V/m$

(d) Energy Density $= \dfrac{1}{2}\varepsilon_0 E^2 = \dfrac{1}{2}(8.85 \times 10^{-12} C^2/N-m^2)(3 \times 10^6 V/m)^2$
Energy Density = 39.8 J/m^3

☐☐ Consider the figure. A 20 µF capacitor is hooked to a 12 V battery and charged. It is then disconnected from the battery and hooked across the terminals of an uncharged 8 µf capacitor. Find (a) the charge on the 20 µF capacitor after being charged by the 12 V battery, (b) the potential difference across the plates of each capacitor after they have been hooked together, and (c) the charge on the 20 µF and 8 µF capacitors after the two are hooked together.

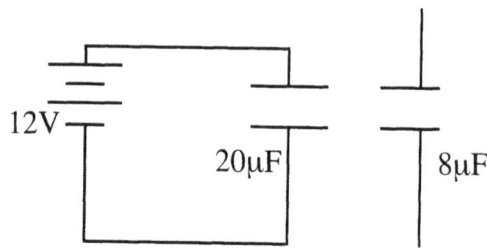

(a) $C = \dfrac{Q}{V} \Rightarrow Q = CV = (20\ \mu F)(12\ V) = 240\ \mu C$

(b) After being disconnected from the battery, and being hooked to the 8 μF capacitor, these two capacitors act as if they were in parallel. Thus,
$$C_{EQ} = C_1 + C_2 = 28\mu F \Rightarrow V = \dfrac{Q}{C} \Rightarrow V = \dfrac{240\mu C}{28\mu F}$$

$V_{EQ} = 8.57$ V equals the potential difference across the plates of each capacitor.

(c) $Q_1 = C_1V_1 = (20\ \mu F)(8.57\ V) = 171\ \mu C$
$Q_2 = C_2V_2 = (8\ \mu F)(8.57\ V) = 69\ \mu C$

Note that $Q_{C1} + Q_{C2} = Q_{total} = 240\ \mu C$

□□ A 4 μF, 6 μF, and 12 μF capacitor are hooked in series using a 24 V battery, as in the figure. Find (a) the equivalent capacitor, (b) the charge on each capacitor, and (c) the potential difference across the plates of each capacitor.

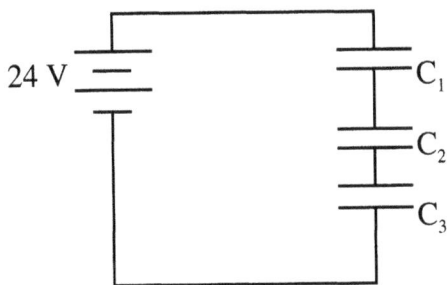

(a) $\dfrac{1}{C_{EQ}} = \dfrac{1}{C_1} + \dfrac{1}{C_2} + \dfrac{1}{C_3} = \dfrac{1}{4\mu F} + \dfrac{1}{6\mu F} + \dfrac{1}{12\mu F}$

$C_{EQ} = 2\ \mu F$

(b) $Q = CV \Rightarrow Q_{EQ} = C_{EQ}V = (2\ \mu F)(24\ V) = 48\ \mu C$
Since capacitors hooked in series have the same charge, $Q_1 = Q_2 = Q_3 = 48\ \mu C$

(c) $V_1 = \dfrac{Q_1}{C_1} = \dfrac{48\mu C}{4\mu F} = 12V$

$V_2 = \dfrac{Q_2}{C_2} = \dfrac{48\mu C}{6\mu F} = 8V$

$V_3 = \dfrac{Q_3}{C_3} = \dfrac{48\mu C}{12\mu F} = 4V$

Note: The potential differences should sum to the applied potential difference of 24V.

❏❏ A defibrillator unit consists of two paddles hooked to a capacitor. The energy from the capacitor is used to shock the heart into beating. If a defibrillator unit consists of a 170 µF capacitor charged to 2000 V, what energy is stored in the capacitor?

$U_C = (1/2) CV^2 = (1/2) (170 \times 10^{-6} F)(2000 V)^2$
$U_C = 340 J$

❏❏ A parallel plate air capacitor is connected to a 12 V battery and then charged. The capacitor is then disconnected from the battery and a material with dielectric constant of 2.8 is inserted between the plates. Find the new potential difference across the plates.

First, $C = \dfrac{Q}{V} \Rightarrow Q = CV$, and $K = \dfrac{C}{C_0}$. Thus, $Q = C_0 V_0 = CV$. Since $C = KC_0$,

$C_0(V_0) = K(C_0)V \Rightarrow C_0(12\ V) = 2.8(C_0)V$

$V = 4.29\ V$

So that the new potential difference is 4.29 V.

❏❏ A 200 µF, 2,000 V capacitor acts as a backup power supply and is charged from a 2,000 V source. How long would the energy stored in the capacitor operate a 7.5 watt radio?

$U_C = (1/2)CV^2 = (1/2)(200 \times 10^{-6} F)(2000 V)^2 = 400 J.$

Since Power $= \dfrac{\Delta E}{t} \Rightarrow t = \dfrac{\Delta E}{Power} = \dfrac{400 J}{7.5 J/s} = 53.3s$

❏❏ The space between a capacitor is filled with two dielectrics of equal size as in the figure. Find the capacitance in terms of K_1, and K_2 and C_0, the capacitance with air between the plates.

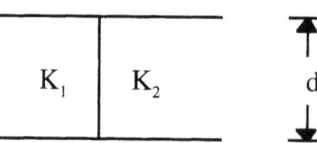

Essentially, this is two capacitors in parallel. Thus,

$C = \dfrac{K_1 C_0}{2} + \dfrac{K_2 C_0}{2} = \dfrac{C_0}{2}(K_1 + K_2)$

⬜⬜ **Two capacitors of value 4 µF and 6 µF are connected in series with a 20V battery. Find (a) the equivalent capacitance, (b) the total energy stored, and (c) the charge on each capacitor.**

(a) For capacitors in series, $\dfrac{1}{C_{EQ}} = \dfrac{1}{C_1} + \dfrac{1}{C_2} = \dfrac{1}{4\mu F} + \dfrac{1}{6\mu F}$

$C_{EQ} = 2.4\ \mu F$

(b) $U_C = (1/2)C_{EQ}V^2 = (1/2)(2.4 \times 10^{-6}\ F)(20\ V)^2$

$U_C = 4.8 \times 10^{-4}$ J is the total energy stored.

(c) Since capacitors in series have the same charge, the charge on each will be the same as the charge on the equivalent capacitor.

$Q_{EQ} = C_{EQ}V = (2.4\ \mu F)20\ V = 48\ \mu C$, so that $Q_1 = Q_2 = 48\ \mu C$

⬜⬜ **When two capacitors are hooked in parallel, their equivalent capacitance is 20 µF. When the capacitors are hooked in series, their equivalent capacitance is 4.2 µF. Find the values of the capacitors.**

For the parallel connection $C_{EQ} = C_1 + C_2 = 20\ \mu F$

For the series connection $\dfrac{1}{C_{EQ}} = \dfrac{1}{C_1} + \dfrac{1}{C_2} = 4.2\mu F$

Thus, $C_1 = 20\ \mu F - C_2$ so that $\dfrac{1}{20\mu F - C_2} + \dfrac{1}{C_2} = \dfrac{1}{4.2\mu F}$

or $\dfrac{20\mu F}{(C_2)20\mu F - C_2^2} = \dfrac{1}{4.2\mu F}$

$84 = 20C_2 - C_2^2$ so that $C_2 = 6\mu F \Rightarrow C_1 = 14\ \mu F$

or $C_2 = 14\ \mu F \Rightarrow C_1 = 6\mu F$

⬜⬜ **A circular capacitor has paper between the parallel plates of radius 5 cm separated by 0.2 mm. Given that the dielectric strength of paper is 12 × 10⁶ V/m, and that the dielectric constant is 3.5, find (a) the capacitance, and (b) the maximum voltage that can be applied to the plates of this capacitor.**

(a) $C = \dfrac{K\varepsilon_0 A}{d} = \dfrac{3.5(8.85 \times 10^{-12}\ C^2/N-m^2)\pi(.05m)^2}{2 \times 10^{-4}\ m}$

$C = 1.22 \times 10^{-9}$ F = 1.22 nF

(b) Dielectric strength = $E_{max} = \dfrac{V_{max}}{d}$.

Thus, $V_{max} = (E_{max})d = 12 \times 10^6$ V/m$(2 \times 10^{-4}$ m$)$

$V_{max} = 2400$ V

☐☐ **A 40 µF capacitor has an initial potential difference of 400 V. The capacitor is completely discharged in 0.6 ms. What average power has been delivered by the capacitor?**

$$\text{Average Power} = \frac{\text{work done}}{\text{time}} = \frac{\text{change in energy}}{\text{time}}$$

Thus, $P_{av} = \frac{U_C}{\text{time}} = \frac{\frac{1}{2}CV^2}{\text{time}} = \frac{\frac{1}{2}(40 \times 10^{-6} F)(400V)^2}{0.6 \times 10^{-3} s}$

$P_{av} = 5330$ watts

☐☐ **A parallel plate capacitor has paper (dielectric constant 3.5, dielectric strength of 12 x 10⁶ V/m) between the plates. Find (a) the area of one of the plates if the capacitance value is 6 µF, and (b) the thickness of the paper if the maximum operating potential is 5,000 volts.**

(a) Since $E_{max} = \frac{V_{max}}{d} \Rightarrow d = \frac{5000V}{12 \times 10^6 V/m} = 0.417mm$

Hence, $C = \frac{K\varepsilon_0 A}{d} \Rightarrow \frac{Cd}{K\varepsilon_0} = A = \frac{(6 \times 10^{-6} F)(4.17 \times 10^{-4} m)}{(3.5)(8.85 \times 10^{-12} C^2/N-m^2)}$

A = 80.8 m

(b) From (a) above, d = 4.17 x 10⁻⁴ m

☐☐ **Find the capacitance between points A and B of the figure. All capacitors have value of 20 µF.**

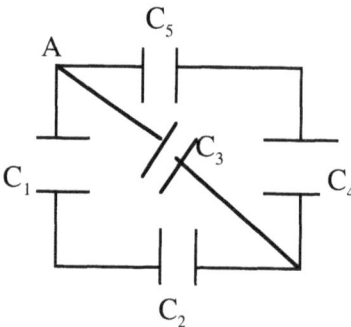

Capacitors C_1 and C_2 are in series, as are C_5 and C_4. Thus,

$$\frac{1}{C_{EQ}} = \frac{1}{C_1} + \frac{1}{C_2} = \frac{1}{20\mu F} + \frac{1}{20\mu F} \Rightarrow C_{EQ} = 10\mu F$$

Resketching the circuit, it consists of three capacitors in parallel with each other. Thus

C_{AB} = 10 µF + 20 µF + 10 µF = 40 µf

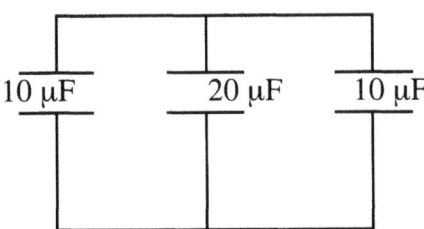

☐☐ A 20 PF (P = 1 x 10⁻¹²) air capacitor (Dielectric strength of air is 3 x 10⁶ V/m) has plates of area 0.05 m². Find (a) that potentials difference for which dielectric breakdown occurs, and (b) the charge on the plates of the capacitor when dielectric breakdown occurs.

(a) $C = \dfrac{\varepsilon_0 A}{d} \Rightarrow d = \dfrac{\varepsilon_0 A}{C} = \dfrac{(8.85 \times 10^{-12} C^2/N-m^2)(0.05 m^2)}{20 \times 10^{-12} F} = 2.21 \times 10^{-2} m$

Dielectric strength $= E_{max} = \dfrac{V_{max}}{d} \Rightarrow V_{max} = E_{max} d$

$V_{max} = (3 \times 10^6 V/m)(2.21 \times 10^{-2} m) = 66,400 V$

(b) $C = \dfrac{Q}{V} \Rightarrow Q = CV$

$Q = (20 \times 10^{-12} F)(66,400 V)$

$Q = 1.33 \, \mu C$

CURRENT, RESISTANCE AND SOURCES OF EMF

THE FUNDAMENTAL IDEAS ARE THE FOLLOWING:

- Current (on the average) equals charge/time, i.e., $I = Q/t$. Current is measured with an ammeter, symbol

- Ohm's Law says that $V = IR$, where R is said to be Resistance, which is measured in ohms, symbol Ω, however, not all materials obey Ohm's Law.

- The Franklin convention says that current flows from positive to negative outside the source (which means it flows from negative to positive through the source).

- When charge (or current) flows through a resistor, heat is produced.

- It takes two things to specify a resistor: (1) its resistance value, and (2) its power rating. Thus a 10 Ω, 5 watt resistor means that the resistor has a resistance of 10 Ω and is capable of dissipating 5 Joules per second of heat energy. If the power rating of the resistor is exceeded, it will be destroyed.

- Electrical energy is sold commercially in units of Kilowatt-hours, kWh. 1 kWh is the same as 3.6×10^6 J.

- Joule's Law of heating says that the heat produced in a resistor is found by using the definition of potential difference, i.e. $V = \frac{work}{Q} \Rightarrow work = \Delta Energy = VQ = VIt$

    or $\Delta Energy = I^2Rt$
    or $\Delta Energy = \frac{V^2}{R}t$.

- Since Power = $\frac{work\ done}{time} = \frac{\Delta Energy}{time}$
  This says that $P = VI = I^2R = V^2/R$ where P is electrical power.

- The resistance of a length of wire is given by $R = \rho \frac{l}{A}$, where l is length, A is cross-sectional area, and ρ is called resistivity. Note that the units of ρ are Ω-m. The resistivity of conductors is small, while the resistivity of insulators is high.

- The change in resistance of a resistor, because it experiences a change of temperature, is found by $\Delta R = \alpha R_0 \Delta t$, where R_0 is original resistance, Δt is change in temperature in °C, and α is called the temperature coefficient of resistivity. Note that most materials have a resistance that increases with an increase in temperature, but carbon is a material that experiences a decrease in resistance as the temperature increases.

- Series means only one path for current to follow. Thus for resistors hooked in series as in the figure, the equivalent resistance R_{EQ} is found by $R_{EQ} = R_1 + R_2 + R_3$.

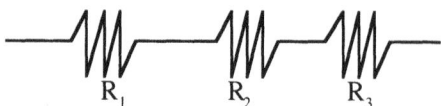

- Parallel means more than one path for current to follow. Thus, for resistors hooked in parallel as in the figure, the equivalent resistance R_{EQ} is found by
$$\frac{1}{R_{eq}} = \frac{1}{R_1} + \frac{1}{R_2} + \frac{1}{R_3}$$

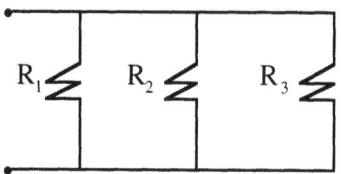

- Resistors can be hooked in a series-parallel combination.

- A source of electromotive force, emf, is a device that causes charge (current) to flow. A battery is an example of a source of emf. Thus, when charge flows through a resistor, it loses energy. When charge flows through a battery it gains energy.
  $E = \frac{work}{Q}$ is the definition of emf. When the charge flows through a source of emf, it gains energy in the amount of EQ.

- An ideal battery has no internal resistance.

- The schematic symbol for a battery is as in the figure, with the long line representing the positive terminal of the battery.

SOLVED PROBLEMS

❏❏ **A battery maintains a current of 25 A for a period of 15 minutes. How much charge flows through the battery?**

$I = \frac{Q}{t} \Rightarrow Q = It = (25 \text{ C/s})900\text{S} = 2.25 \times 10^4 \text{C}$.

❑❑ The resistor in the figure is rated at 25 Ω, 4 watts. Find (a) the potential drop across the resistor, and (b) whether or not the resistor will burn up.

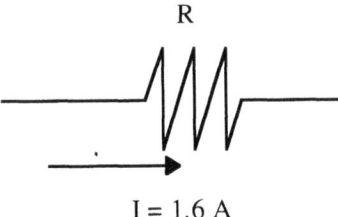

I = 1.6 A

(a) V = IR = (1.6 A)(25 R) = 40V

(b) P = VI = (40V)(1.6A) = 64 watts
    = I²R = (1.6A)² (25 Ω) = 64 watts
    = V²/R = (40 V)²/(25 Ω) = 64 watts
Since, 64 watts > 4 watts, the resistor will burn up.

❑❑ A copper wire has a resistance of 5 Ω. It is drawn through a die so that its length is doubled, and its cross-sectional area is reduced (total volume remains constant). What is its new resistance?

$R_0 = \dfrac{\rho l_0}{A_0} = 5\Omega$, $\quad l = 2l_0$, and $A = \dfrac{A_0}{2}$

Thus, $R = \dfrac{\rho l}{A} = \dfrac{\rho(2l_0)}{\dfrac{A_0}{2}} \Rightarrow R = \dfrac{4\rho l_0}{A} = 4R_0$, so that the new resistance is 20 Ω.

❑❑ If the copper wire of the previous problem had 240 V applied to it, what (a) is the current through the wire and (b) what heat energy per unit time would appear in the wire?

(a) V = IR ⇒ 240 V = I(45 Ω)
    I = 48 A through the wire.

(b) P = VI = (240 V)(48 A) = 1.15 x 10⁴J/s, so that each second the current flowing through the wire produces 1.15 x 10⁴ J of heat.

❑❑ A resistor is used to heat water. How long would it take a 800 watt heating element to raise 0.6 kg of water from 0 °C to 80 °C, assuming that no heat is lost to the surround? (Specific heat of water is 4186 J/kg)

Power = $\dfrac{\Delta Energy}{time}$ ⇒ time = $\dfrac{\Delta Energy}{Power}$. In this case electrical energy is changed to heat energy, thus, heat energy = mCΔt.

mC Δt = (0.6 kg)(4186 J/kg°C)(80 °C0 = 2.01 x 10⁵ J, so that time = $\dfrac{2.01 \times 10^5 J}{800 watts}$ = 251s.

or time = 4.19 minutes.

◻◻ Given that the resistivity of copper is 1.72 x 10^{-8} Ω-m, what would be the resistance of 16 km of copper wire of cross section 6 x 10^{-5} m²?

$$R = \frac{\rho l}{A} = \frac{(1.72 \times 10^{-8} \Omega - m)(16,000m)}{6 \times 10^{-5} m^2}$$

R = 4.59 Ω

◻◻ A potential difference of 12 V is applied across a resistance of 4 Ω for a period of 6 minutes. Find (a) the current through the resistor, (b) the heat generated in the resistor, (c) what energy per unit time is dissipated in the resistor and (d) what charge flows through the resistor in 5 minutes.

(a) V = IR ⇒ I = V/R = (12 V)/4Ω = 3A

(b) Heat = VIt = (12 V)(3A)(360s)
Heat = 1.3 x 10^4 J

(c) $P = \frac{\Delta Energy}{time} = VI = 36 J/s$

(d) $I = \frac{Q}{t} \Rightarrow Q = It = (3A)300s$

Q = 900 C

◻◻ A resistance thermometer (used to determine the temperature of an oven) has a coefficient of resistivity of 0.0037/ °C. The thermometer has a resistance of 100 Ω at 10 °C. Find the temperature of the oven when the coil has a resistance of 325 Ω.

$$\Delta R = \alpha R_0 \Delta t \Rightarrow \frac{\Delta R}{\alpha R_0} = \Delta t = \frac{225 \Omega}{(0.0037/°C)(100\Omega)}$$

Δt = 608 °C. Thus, the temperature of the oven is 618 °C.

◻◻ A 100 watt light bulb has 120 V applied to it. Find (a) the resistance of the bulb, (b) the current through it, (c) the charge through the light bulb each second, and (d) the number of electrons that pass through the bulb in each second.

(a) Since $P = \frac{V^2}{R} \Rightarrow R = \frac{V^2}{P} = \frac{(120V)^2}{100 watt}$

R = 144 Ω

(b) $V = IR \Rightarrow I = \frac{V}{R} = \frac{120V}{144\Omega} = 0.833A$

(c) Since $I = \frac{Q}{t} \Rightarrow Q = It = (.833 C/s)(1s)$

Q = .833C through the lamp each second.

(d) Since $Q = nq \Rightarrow n = \frac{-Q}{q} = \frac{-.833C}{-1.6 \times 10^{-19} C/e^-}$

R = 5.21 x 10^{18} electrons per second. Note: Since the electrons are the charges moving, the charge that passes through the filament is -.

☐☐ **A 12 V battery has a current of 6A flowing through it. How much energy does each coulomb of charge receive when it flows through the battery?**

$E = \dfrac{work}{Q}$ ⇒ work = EQ = (12J/C)(1C). Thus, work done = change in energy = 12 J.

☐☐ **For the battery of the previous problem, how much charge flows through the battery in 1 hour?**

$I = \dfrac{Q}{t}$ ⇒ Q = It = (6C/s)(360 s)

Q = 2.16 x 10^4C

☐☐ **Find the resistance between points A and B of the figure. Each resistor has a resistance of 10 Ω.**

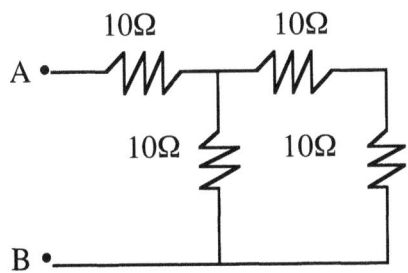

Note that this circuit reduces to

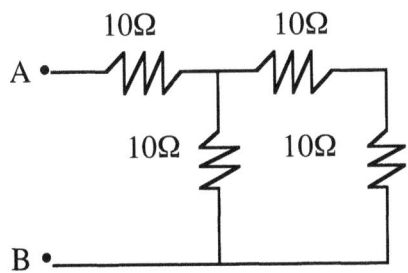

Since the 20 Ω and 10 Ω resistors are in parallel, $\dfrac{1}{R_{EQ}} = \dfrac{1}{10\Omega} + \dfrac{1}{20\Omega}$ = 6.67 Ω,

so that

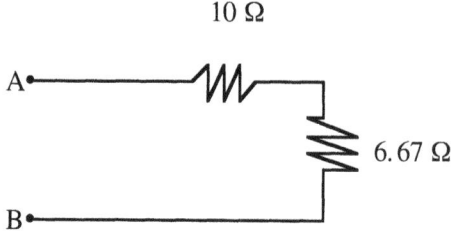

Finally $R_{AB} = 16.7\ \Omega$, since the 10 Ω and 6.67 Ω resistor are in series.

☐☐ **Which has a higher resistance, a 300 watt, 120 V light bulb or a 60 watt, 120 V light bulb?**

$P = \dfrac{V^2}{R} \Rightarrow R = \dfrac{V^2}{P}$. Thus, $R_{300watt} = \dfrac{(120V)^2}{300 watts} = 48\Omega$

$R_{60watt} = \dfrac{(120V)^2}{60 watts} = 240\Omega$.

Thus, the 60 watt bulb has a higher resistance.

☐☐ **A toaster uses nichrome wire for the heating element (α for nichrome is 0.0004/ °C). When the toaster is turned on at 25 °, the initial current is 1.5 A. After the toaster has warmed up, the current falls to 1.25 A. Given that the toaster is plugged into a 120 V wall socket, find the temperature of the heating element after it has warmed up.**

$\Delta R = \alpha R_0 \Delta t$, and $V = IR$.
Hence, $R_0 = \dfrac{V}{I_0} = \dfrac{120V}{1.5A} = 80\Omega$, the initial resistance.

$R = \dfrac{V}{I} = \dfrac{120V}{1.25A} = 96\Omega$, the final resistance.

Thus, $16\ \Omega = (0.0004/°C)(80\Omega)\Delta t \Rightarrow \Delta t = 500°C$

Finally, the temperature of the heating element after it has warmed up is 525° C.

☐☐ **For the toaster of the previous problem, if electricity costs 10 cents /kWh, how much does it cost to run the toaster for 20 minutes, assuming that the current is 1.25 H for this time interval?**

$P = VI = (120\ V)(1.25\ A) = 150\ W$. Since Power $= \dfrac{\Delta E}{time} \Rightarrow \Delta E = (Power)(time)$

$\Rightarrow \Delta E = (150W)(\tfrac{1}{3}h) = 50Wh = 0.05kWh$

Since electricity costs 10 cents /kWh, cost $= 0.05kWh(.10/kWh) = 0.5$ cents. It cost 1/2 cent to run the toaster for 20 minutes.

❏❏ An electric furnace runs 8 hours per day to heat a house. The heating element has a resistance of 5.1 Ω, and has 220 V applied to it. Find the cost of running this furnace for the month of December if electricity costs 10 cents /kWh.

$$P = \frac{\Delta E}{time} \Rightarrow \Delta E = Pt = \left(\frac{V^2}{R}\right)t = \frac{(220V)^2}{5.1\Omega}(8h)$$

Thus, $\Delta E = 7.59 \times 10^4$ W = 75.9 kWh.

Thus, for one day, cost = 75.9 kWh(10cents/kWh)
cost = 759 cents/day = $7.59/day

For December (31 days), the cost is $235.29.

❏❏ Find the resistance between A and B is the figure. Each resistor is 15 Ω.

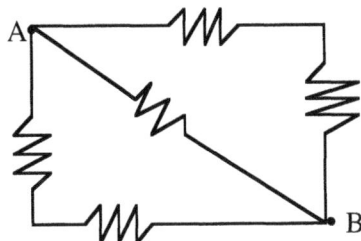

Resketching the diagram, and making use of the fact that resistors in series have an equivalent resistance given by $R_1 + R_2 + R_3 + ...$, one has the figure below:

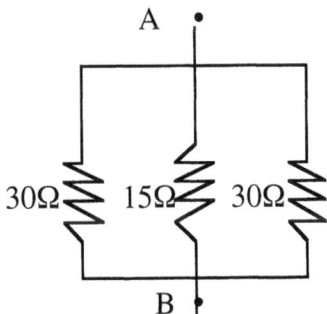

For resistors in parallel, $\frac{1}{R_{EQ}} = \frac{1}{R_1} + \frac{1}{R_2} + \frac{1}{R_3} + ...$

Thus, $\frac{1}{R_{EQ}} = \frac{1}{30\Omega} + \frac{1}{15\Omega} + \frac{1}{30\Omega} = \frac{4}{30\Omega}$

Hence, $R_{EQ} = R_{AB} = 7.5\ \Omega$

◻◻ **For the previous problem if 120 V is applied to points A and b, what current flows?**

$V = IR \Rightarrow I = \dfrac{V}{R} = \dfrac{120V}{7.5\Omega} = 16A$. Thus, 16 A of current would flow between points A and B if 120 V is applied to points A and B.

◻◻ **A wire has an initial resistance of 35 Ω at 20 °C. In boiling water, its resistance is 47.6 Ω. Find the temperature when the resistance of the wire is 36.9 Ω.**

$\Delta R = \alpha R_0 \Delta t \Rightarrow (47.6\Omega - 35\Omega) = \alpha(35\Omega)80°C$

so that the coefficient of resistivity of the wire is 4.5 x 10^{-3}/°C. Then, for the second situation

$\Delta R = \alpha R_0 \Delta t \Rightarrow \Delta t = \dfrac{\Delta R}{\alpha R_0} = \dfrac{(36.9-35)\Omega}{(4.5x10^{-3}/°C)35\Omega}$

Δt = 12.1 °C. The temperature when the resistance of the wire is 36.9 Ω is 32.1 °C.

◻◻ **A thermistor has a temperature coefficient of resistivity of -0.06 / °C. The resistance of the thermistor decreases to 80 % of its value at 37 °C. What is the temperature of the thermistor?**

ΔR = αR₀Δt
(0.8 R₀ - R₀) = -.06/°C(R₀)Δt
Δt = 3.33 °C.
Thus, the temperature of the thermistor is 40.3°.
Note that a material with a negative temperature coefficient of resistivity experiences a decrease in resistance as its temperature increases.

◻◻ **A cauterizer (used to stop bleeding in surgery) puts out 3 mA at 15,000 V. Find (a) its power output and (b) the resistance of the tissue of its path.**

(a) P = VI = (0.003A)(15,000 V) = 45 W

(b) $R = \dfrac{V}{I} = \dfrac{15,000V}{.003A} = 5x10^6 \Omega$

◻◻ **In 1955, GM switched from a 6 V battery system to a 12 V battery system. Why would the cables from the battery for the 12 V system be smaller than the cables from the battery for the 6 V system?**

P = VI. Thus, if the potential difference goes up, the current requirement goes down for a constant power. Hence, the wires that carry the current can be smaller for the 12 V system.

◻◻ **14 gauge copper wire (diameter 1.628 mm) is designed to carry a maximum current of 15 A. Given that the resistivity of copper is 1.72 x 10^{-8} Ω-m, find (a) the resistance of 100 m of 14 gauge wire, and (b) the potential difference across 100 m of the wire when it is carrying 15 A of current.**

(a) $R = \dfrac{\rho l}{A} = \dfrac{(1.72x10^{-8}\Omega - m)(100m)}{\pi(8.14x10^{-4}m)^2} = 0.827\Omega$

(b) V = IR = (15 A)(0.827 Ω) = 12.4 V

⬜⬜ A power transmission line transmits 200,000 V and is held in place with ceramic insulators that have a resistance of 1.1 x 10⁹ Ω. What current flows through the insulator?

$$V = IR \Rightarrow I = \frac{V}{R} = \frac{200,000V}{1.1 \times 10^9 \Omega}$$

I = 0.182 mA

⬜⬜ An insulated container holds 0.400 kg of water at 25 °C. An immersion heater raises the temperature of the water to the boiling point in 4 minutes. (Specific heat of water is 4186 J/kg °C). Find (a) the power rating of the heater, and (b) how much it cost to heat the water, assuming that electricity costs 10 cents /kWh and that all of the heat generated by the heater heats the water.

(a) Power = $\frac{\Delta Energy}{time} = \frac{mc\Delta t}{time} = \frac{(0.4kg)(4186 J/kg°C)(75°)}{240s}$

Power = 523 W

(b) $\Delta Energy$ = (Power)(time) = (.523kW)($\frac{1}{15}$h) = 0.0349 kWh.

Since cost = 0.0349kWh(10 cents/kWh) = 0.349 cents, it costs about 0.35 cents to heat the water.

⬜⬜ Given that the temperature coefficient of resistivity for aluminum is 3.9 x 10⁻³/ °C at 20°C, to what temperature must we raise the wire for its resistance to double (ignore changes in dimension because of heating).

$\Delta R = \alpha R_0 \Delta t$
$(2R_0 - R_0) = (3.9 \times 10^{-3}/°C)R_0 \Delta t \Rightarrow \Delta t = 256°C$

Thus, if the temperature of the wire is raised to 276 °C, its resistance will be doubled.

⬜⬜ For the figure, what would (a) the voltmeter read, (b) what would the ammeter read, and (c) would the resistor burn up?

E = 12 V

R = 16Ω, 4watts

(a) I = $\frac{V}{R} = \frac{12V}{16\Omega}$ = 0.75A

(b) V = 12 V (because that is what the battery applies to the resistor)

(c) P = VI = (12 V)(.75 A) = 9 watts.
   Since, 9 watts > 4 watts, the resistor will burn up.

☐☐ **For the previous figure, if the Franklin convention is followed, which end of the resistor is positive?**

The right end, because the Franklin convention says that current flows from + → − outside the source.

DIRECT CURRENT CIRCUITS

THE FUNDAMENTAL IDEAS ARE THE FOLLOWING:

- A battery is a source of emf, E.

- For a battery that is discharging, the potential difference at the terminals, V_t, is found by $V_t = E - Ir$, where I is current and r is the internal resistance of the battery.

- For a battery that is being charged, current flows through the battery in oppostion to the Franklin convention (electrical energy is being changed into chemical energy). In this case, the terminal potential difference V_t is given by $V_t = E + Ir$, where r is the internal resistance of the battery.

- If batteries are hooked in series with unlike terminals hooked together, that is a series-aiding connection where the emf of the combination is the sum of the individual emf/s, or $E_c = E_1 + E_2$

- If batteries are looked in series with like terminals hooked together, that is a series-opposing connection where the emf of the combination is the difference of the individual emf's or $E_C = E_1 - E_2$.

- For a series circuit there are two important ideas:
  a) The current everywhere is the same.
  b) The applied potential difference equals the sum of the potential differences across each element.

- For a parallel circuit there are two important ideas:
  a) The potential difference across each branch is the same as the applied potential difference.
  b) The total current is the sum of the current in each branch.

- To solve a circuit means to find the currents that flow, because if the current is known one can find the potential difference across resistors, terminal potential difference of batteries, etc.

- Ohm's Law for a series circuit says that $I = \dfrac{E_1 + E_2 + ...}{R_1 + R_2 + ...} = \dfrac{\Sigma E's}{\Sigma R's}$

- Kirchoff's two laws are:
  a) The sum of the currents into a point equal the sum of the currents out of that point, or, $\Sigma I_{in} = \Sigma I_{out}$ of a point.

  b) The sum of the potential rises equals the sum of the potential drops around any closed loop, or, $\Sigma P.R.'s = \Sigma P.D.'s$ around any closed loop.

  Note if one travels from $+ \rightarrow -$, it is a potential drop, while if one travels from $- \rightarrow +$ it a potential rise.

- The five step recipe to apply Kirchoff's two laws is the following:
  a) Assign a direction and a symbol to the flow of current in each branch.

  b) Place appropriate + and – signs on resistors.

  c) Apply Kirchoff's 1st Law enough times so that each current appears at least once.

d) Apply Kirchoff's 2nd Law enough times so that each current appears at least once.

e) Solve equations.

Note that if one or more of the currents are negative, that simply means that current flows in a direction opposite to the one chosen.

- A D'arsonval galvanometer is a device that is a very sensitive ammeter.

- To convert a galvanometer to a voltmeter of any range, a resistor is hooked in series with the galvanometer.

- To convert a galvanometer to an ammeter of any range, a resistor (called a shunt resistor) is hooked in parallel with the galvanometer.

SOLVED PROBLEMS

❐❐ Consider the circuit in the figure. Find (a) the current that flows, (b) the voltage drops across the 4 Ω resistor, and (c) the terminal potential difference of the battery. [Note that the dots represent the terminal of the battery.]

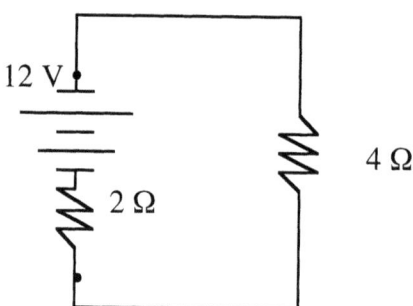

(a) Since this is a series circuit, Ohm's Law for a series circuit is applicable. Thus,
$$I = \frac{\Sigma E's}{\Sigma R's} = \frac{12V}{6\Omega} = 2A$$

(b) The voltage drop across the 4 Ω resistor is found from Ohm's Law for a series resistor; V = IR = (2A)(4Ω) = 8V.

(c) The terminal potential difference of the battery is found by V_t = E-Ir. Hence, V_t = 12V - 2A (2 Ω) = 8V. Note that the terminal potential difference of the battery is the potential difference applied to the external circuit.

❏❏ Consider the figure. If a 15 V potential difference is applied to the battery, what current would flow if the + terminal of the 15 V source is applied to the + terminal of the battery?

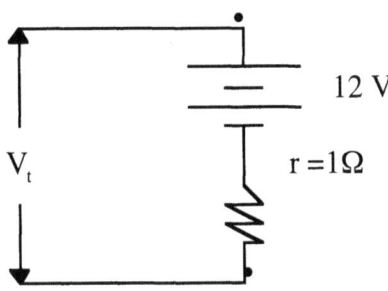

Here, the battery is being charged. Thus, $V_t = E + Ir \Rightarrow 15\text{ V} = 12\text{ V} + I(1\text{ }\Omega)$
$\dfrac{3V}{1\Omega} = I = 3\text{A}$

❏❏ For the circuit in the figure:
(a) Are the batteries hooked series-aiding or series opposing?
(b) Which battery is being charged?
(c) What current flows?
(d) What is the terminal potential difference of the 12 V battery?
(e) What is the terminal potential difference of the 3 V battery?

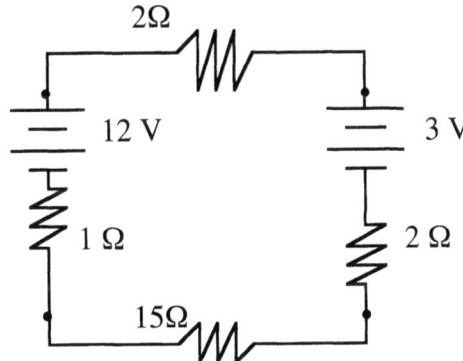

(a) The batteries are hooked series-opposing because like terminal are hooked together.

(b) The 3 V battery is being charged because the 12 V battery is forcing current through it backwards.

(c) $I = \dfrac{\Sigma E's}{\Sigma R's} = \dfrac{12V - 3V}{1\Omega + 2\Omega + 2\Omega + 15\Omega} = \dfrac{9V}{20\Omega}$

$I = 0.45\text{ A}$

(d) Since the 12 V battery is discharging, $V_t = E - Ir = 12\text{ V} - (0.45\text{ A})(1\text{ }\Omega) = 11.55\text{ V}$

(e) Since the 3 V battery is being charged, $V_t = E + Ir = 3\text{ V} + (0.45\text{ A})(2\text{ }\Omega) = 3.9\text{ V}$

❑❑ Imagine that your car is equipped with a 12 V battery and a voltmeter. If you glance at the voltmeter while driving and it reads 11 V, should you be concerned?

Your battery is discharging, i.e., it is providing the current to run the car, and it is not being charged. Thus, you should be concerned, because the battery can't provide the current to run the car's electrical system for an unlimited time.

❑❑ Batteries are sometimes rated in ampere-hours, Ah. This means that an 80 Ah battery will provide 1A for 80 h, 2A for 40 H, or 80 A for 1 h. A battery that is rated at 14 mAh is used in a hearing aid. How many days would this battery last if the hearing aid uses 100μA of current? Assume that the hearing aid is on at all times.

Since $I = \frac{Q}{t} \Rightarrow Q = It$ thus, 14×10^{-3}Ah $= (100 \times 10^{-6}$A$)t$,
so that t = 140 h = 5.83 days.

❑❑ Consider the circuit in the figure: Find (a) the potential difference across each resistor, (b) the current through each resistor, and (c) the Ammeter reading.

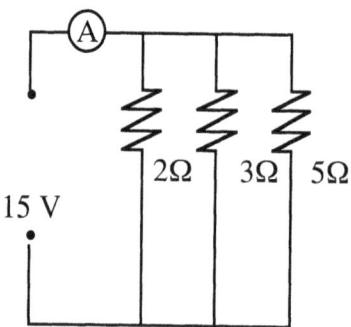

(a) Since the resistors are all in parallel, each resistor has a potential difference of 15 V.

(b) $V = Ir \Rightarrow I = \frac{V}{R}$

Thus, $I_{2\Omega} = \frac{V}{R} = \frac{15V}{2\Omega} = 7.5A$

$I_{3\Omega} = \frac{V}{R} = \frac{15V}{3\Omega} = 5A$

$I_{5\Omega} = \frac{V}{R} = \frac{15V}{5\Omega} = 3A$

(c) The ammeter measures the total current provided by the 15 V source, thus

$I_{Total} = I_{2\Omega} = I_{3\Omega} = I_{5\Omega} = 7.5A + 5A + 3A$
$I_{Total} = 15.5A$

☐☐ Consider the figure: Find (a) what the ammeter reads after a long time, (b) the potential difference across the 8 Ω resistor, (c) the terminal potential difference of the battery, and (d) the charge stored in the capacitor after a long time.

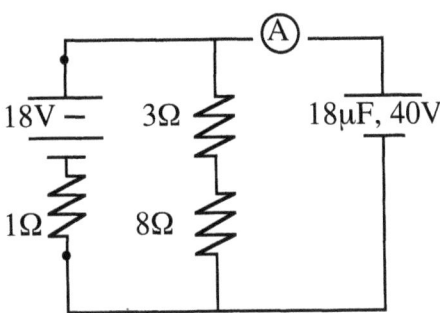

(a) After the capacitor is fully charged, there is no current flow through the ammeter, thus, the ammeter reads 0.

(b) $I = \dfrac{\Sigma E's}{\Sigma R's} = \dfrac{18V}{12\Omega} = 1.5 A$

Thus, $V_{8\Omega} = IR = (1.5\ A)8\Omega = 12$ V

(c) The battery is discharging, thus, $V_t = E - IR = 18$ V $- 1.5\ (1\Omega) = 16.5$ V

(d) For the capacitor, it is in parallel with the 3 Ω and 8 Ω resistor. The equivalent circuit is

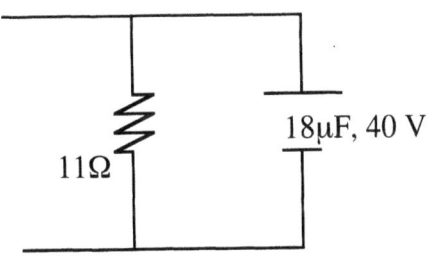

Thus, $V_{11\Omega} = V_C \Rightarrow (I)R = V_C \Rightarrow)1.5\ A)(11\Omega) = V_C = 16.5$ V

Since $C = \dfrac{Q}{V} \Rightarrow Q = CV = (18\mu F)(16.5V) = 297 \mu C$

☐☐ For the figure, (a) apply Kirchoff's First Law at point A, (b) apply Kirchoff's 2nd Law to the left-hand loop, (c) apply Kirchoff's 2nd Law to the outside loop, and (d) find I_1, I_2, and I_3.

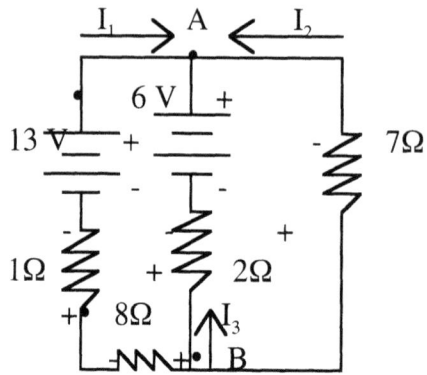

(a) At point A, $\Sigma I_{in} \equiv \Sigma I_{out} \Rightarrow I_1 + I_2 + I_3 = 0$

(b) Around left-hand loop, starting at point B going ccw: ΣPR's $\equiv \Sigma PD$'s
$6 + I_1 + 8I_1 = 2I_3 + 13$

(c) Around the outside loop clockwise starting at point A:
ΣPR's $= \Sigma PD$'s
$7I_2 + 13 = 8I_1 + I_1$

(d) To find I_1, I_2, I_3 there are three equations in three unknowns:
   1) $I_1 + I_2 + I_3 = 0$
   2) $9I_1 - 2I_3 = 7$
   3) $9I_1 - 7I_2 = 13$
by using either a computer or by hand,
   $I_1 = 0.789$ A
   $I_2 = -0.842$ A
   $I_3 = 0.053$ A

The fact that I_3 is - implies that I_3 flows in opposition to the direction chosen. As a check, note that $I_1 + I_2 + I_3 = 0.789$ A $- 0.842$ A $+ 0.053$ A $\cong 0$.

▫▫ **A Galvanometer has a resistance of 23 Ω and a current sensitivity (amount of current for full-scale deflection) of .66 mA (0.00066 A). Find (a) the resistance to change this meter to a 10 V voltmeter and (b) the resistance to change this meter to a 10 A ammeter.**

(a) To change the range of a voltmeter, a resistor is hooked in series. Thus,

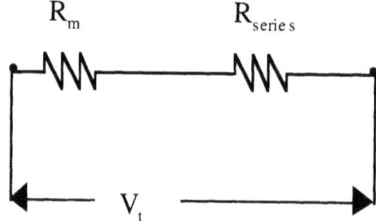

Hence, $V_t = V_m + V_{series} = I_m R_m + I_{series} R_{sereies}$. Since the meter and the resistor are is series, the current through each is the same. Thus, $\dfrac{V_t}{I_m} = R_m + R_{series} = \dfrac{10V}{0.00066A} = 23\Omega + R_{series}$

$\Rightarrow R_{series} \cong 15{,}130 \ \Omega$

Hence, by hooking a 15,130 Ω resistor in series with the meter movement, the meter will indicate 10 V at the full-scale deflection.

(b) To convert the galvanometer to a 10 A ammeter, a shunt (parallel) resistor is used. This resistor shunts around the meter the current over and above the 0.00066 A for full-scale deflection. Thus,

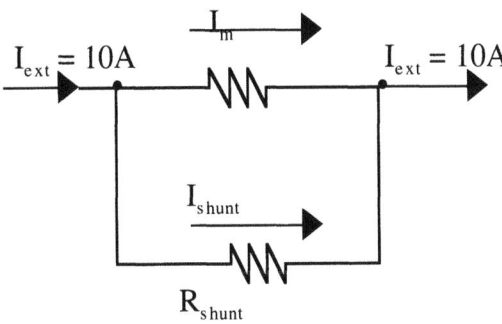

since the meter and the resistor are in parallel, $V_m = V_{shunt} \Rightarrow I_m R_m = I_{shynt} R_{shunt}$

$$\frac{(0.00066A)(23\Omega)}{[10A - 0.00066A]} = R_{shunt}$$

$R_{shunt} = 1.52 \times 10^{-3}$ Ω

Thus, by hooking a 1.52×10^{-3} Ω resistor in parallel with the galvanometer, the meter will indicate 10 A at full-scale deflection.

☐☐ **An ohmmeter is a meter that measures resistance. If a student measured the resistance of a capacitor and (a) the ohmmeter indicated 0 Ω resistance, is the capacitor good or bad? and (b) if the capacitor measured an extremely high (infinite) resistance is the capacitor good or bad?**

(a) Since a capacitor in its usual configuration consists of two conductors separated by an insulator, the resistance should be very high. If the resistance is very low, the capacitor is bad.

(b) This capacitor is good for the reasons stated above.

☐☐ **For the circuit in the figure, find (a) the voltage drop across the 4 Ω resistor, and (b) the terminal potential difference of the 6 V battery.**

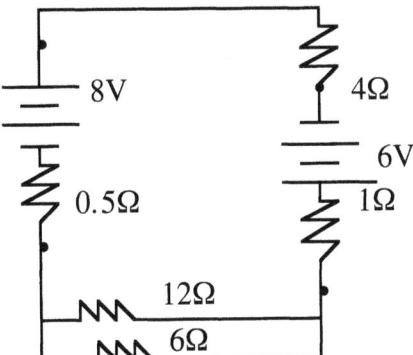

(a) This would be a series circuit if the two resistors in parallel were replaced with their equivalent resistance. Thus, $\frac{1}{R_{eq}} = \frac{1}{12\Omega} + \frac{1}{6\Omega}$, so that $R_{eq} = 4\ \Omega$.

The current that flows is then found by using Ohm's Law for a series circuit,
$$I = \frac{\Sigma E's}{\Sigma R's} = \frac{8V + 6V}{0.5\Omega + 4\Omega + 1\Omega + 4\Omega}$$

$I = 1.47$ A

Finally, $V_{4\Omega} = IR + (1.47A)(4\ \Omega) = 5.88$ V

(b) Since the two batteries are hooked series-aiding, both batteries are providing current to the external circuit, i.e. both batteries are discharging. For the 6 V battery,
$V_t = E - Ir = 6V - 1.47$ A $(1\ \Omega) = 4.53$ V. The terminal potential difference of the 6 V battery is 4.53 V.

☐☐ **Consider the circuit in the figure: Find (a) the current in each branch, and (b) the potential difference across the 6 Ω resistor.**

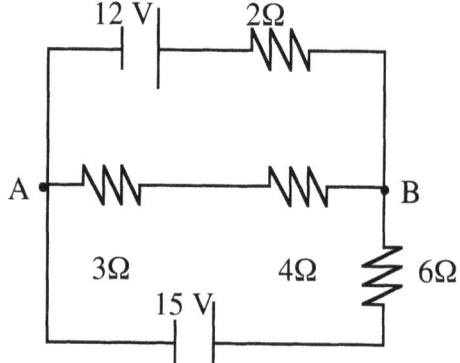

(a) In order to find the current in each branch Kirchoff's laws must be used. Thus, assume that the currents flow as indicated in the figure.

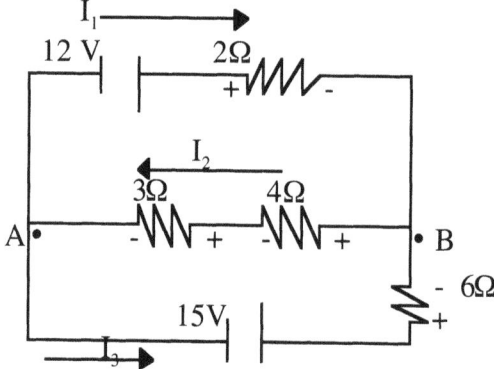

Then, place appropriate + and − signs on resistors. Applying Kirchoff's 1st Law at point A.
$\Sigma I_{in} = \Sigma I_{out}$
$I_2 = I_1 + I_3$
Applying Kirchoff's 2nd Law to the lower loop counter clockwise starting at point B.
Σ PR's = ΣPD's
$15 = 6I_3 + 4I_2 + 3I_2$
Applying Kirchoff's 2nd Law to the outside loop, clockwise starting at point A:
ΣPR's = ΣPD's

$12 + 6I_3 = 2I_1 + 15$

Thus, the three equations for the three unknowns:
$I_1 - I_2 + I_3 = 0$
$7I_2 + 6I_3 = 15$
$2I_1 - 6I_3 = -3$

By using a computer or by hand, the solutions are $I_1 = 0.75$ A; $I_2 = 1.5$ A; $I_3 = 0.75$ A. As a check, $I_2 = I_1 + I_3 \Rightarrow 1.5$ A $= 0.75$ A $+ 0.75$ A.

(b) To find the potential difference across the 6 Ω resistor, $V_{6\Omega} = I_3 R = (0.75$ A$)(6$ Ω$) = 4.5$ V.

☐☐ **A galvanometer has a resistance of 50 Ω and a current sensitivity of 50 μA. Find the resistance to (a) construct a 150 V voltmeter, and (b) construct a 25 A ammeter from this galvanometer.**

(a) To make a 150 V voltmeter a resistor has to be added in series with the meter. Thus,

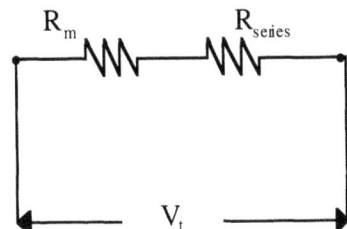

$\Rightarrow V_t = R_m + R_{series}$. since the resistor and the meter are in series with each other, the current through the meter at full-scale deflection will be the current through the resistor. Hence, $\frac{V_t}{I_m} = R_m + R_{series} \Rightarrow \frac{150V}{50 \times 10^{-6} A} = 50\Omega + R_{series}$. Finally, adding a 3MΩ resistor in series with the resistor will change the meter so that full-scale deflection corresponds to 150 V.

(b) To change the meter to a 25 A ammeter, a resistor must be placed in parallel to cause the current over and above 50 μA to be shunted around the meter movement. Hence,

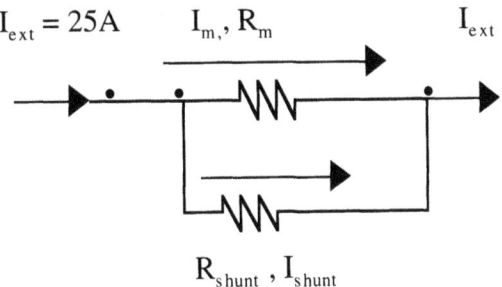

Since the meter and the shunt resistor are in parallel, $V_m = V_{shunt} \Rightarrow I_m R_m = I_{shunt} R_{shunt}$

$\frac{(50 \times 10^{-6} A)(50\Omega)}{25A} = R_{shunt}$

$R_{shunt} = 1 \times 10^{-4}$ Ω. Note that 25 A - 50 μA is very close to 25 A)

Thus, by placing a resistor of value 1×10^{-4} Ω in parallel with the meter movement, then full-scale deflection would correspond to a current of 25 A.

☐☐ A student uses ohmmeter on a lead (piece of connecting wire) as is the figure. The ohmmeter indicates that the lead has an infinite resistance. Is the lead good or bad?

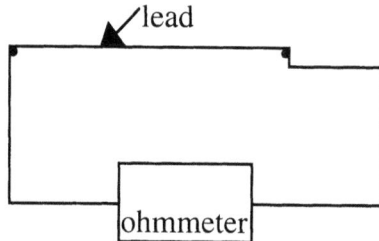

The lead should have a very low resistance to be good, since no current will flow if the resistance is very high. Thus, the lead is bad.

☐☐ A student uses the ohmmeter in the previous problem to check the resistance of a burner element from an electric stove. The resistance of the element is very low, about 10 Ω. Is the burner element good or bad (the stove uses 240 V).

Here V = IR. Since the resistance is low, then there will be current flow, which means that heat will be produced. Thus the burner element is good.

☐☐ Consider the figure. Find (a) the current that flows, and (b) the voltage drop across one of the 15,000 Ω resistors.

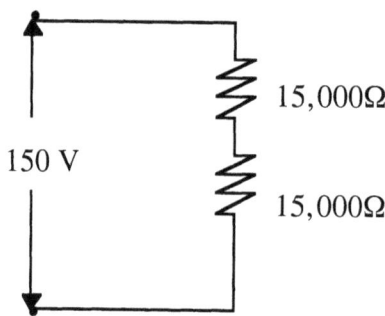

(a) $I = \dfrac{\Sigma E"s}{\Sigma R's} = \dfrac{150V}{30,000 \Omega} = 0.005 A$

(b) V = IR = (0.005 A)(15,000 Ω) = 75 V

◻◻ For the previous problem, suppose that a voltmeter is placed across one of the 15,000 Ω resistors. Find what the voltmeter reads if (a) its resistance is 20,000 Ω and (b) 10,000,000 Ω. This is an example of "loading" a circuit, i.e., the voltmeter affects the reading that it takes.)

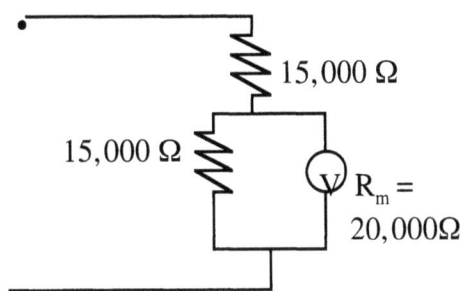

(a) If the voltmeter has a resistance of 20,000 Ω the circuit becomes a 15,000 Ω resistor in series with an equivalent resistance of 8570 Ω. Thus a current of $I = \dfrac{\Sigma E's}{\Sigma R's} = \dfrac{150V}{23570\Omega}$

$I = 6.36 \times 10^{-3}$ A flows.

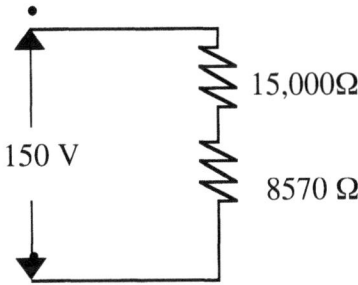

The voltage drop across the 8570 Ω resistor $V = IR = (6.36 \times 10^{-3} \text{ A})(8570 \text{ }\Omega)$
$V = 54.5$ V. This is what the voltmeter reads instead of 75 V.

(b) If the resistance of the meter is 10,000,000 Ω the equivalent resistance of the resistor the voltmeter is across is 14,978 Ω. The voltage drop across the resistor the voltmeter is across is $V = IR = (5.004 \times 10^{-3} \text{ A})(14,978 \text{ }\Omega) = 74.9$ V. Thus, the larger the resistance of a voltmeter the more accurate its reading. The ideal voltmeter has an infinite resistance.

◻◻ Consider the circuit in the figure. Find the current that flows.

$I = \dfrac{\Sigma E's}{\Sigma R's} = \dfrac{12V - 3V}{23\Omega} = 0.391A$

☐☐ For the previous problem, suppose an ammeter with a total resistance of 3 Ω is inserted in the circuit. What does the ammeter read?

Here, $I = \frac{\Sigma E's}{\Sigma R's} = \frac{12V - 3V}{26\Omega} = 0.346 A$. Thus, because of the resistance of the ammeter, the current is read as 0.346 A, rather than its true value of 0.391 A. This is another example of a measurement device that disturbs the very measurement that the instrument is to measure. Note that the ideal ammeter has zero ohm resistance.

☐☐ A battery is to be used in an experimental procedure that requires 200 μA. If the battery is to last six months, what should the Ah (ampere-hour) rating of the battery be?

Since six months equals 365/2 days, and one day is 24 h, six months is 4380 h. Hence, $(200 \times 10^{-6} A)(4380 h) = 0.876$ Ah, the rating of the battery should be at least 0.876 Ah.

☐☐ The electrical system of a car requires 17 A to operate. If the alternator (generator) of a car quits, how long in minutes would an 80 Ah battery operate the car?

80 Ah = (17 A)t ⟹ t = 4.71 h. Since 1 h = 60 min, the car would operate
$t = 4.71 h \left(\frac{60 \min}{1h}\right) \cong 282 \min$.

☐☐ Suppose that two identical batteries are hooked in parallel. What would be the advantage?

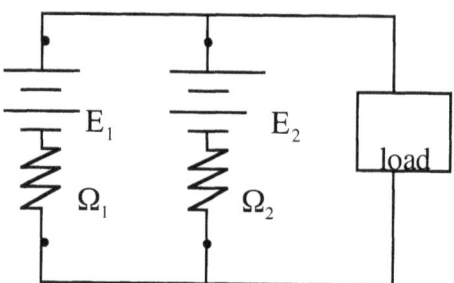

Here the advantage is that the combination can produce more current than can a single battery. Note that the load applied potential difference is the same as that of a single battery.

☐☐ If two batteries are in series, what is the reason if they are hooked (a) series-aiding, and (b) series opposing?

(a) Since for series aiding, the combination emf $E_C = E_1 + E_2$, the reason batteries are hooked series aiding is because it increases the emf.

(b) For series-opposing, the emf of the combination $E_C = E_1 - E_2$. One reason for hooking batteries in this fashion is to use the larger battery to charge the smaller.

MAGNETIC FIELDS AND FORCES

THE FUNDAMENTAL IDEAS ARE THE FOLLOWING:

- Any region in which magnetic forces are observed is a region of magnetic field.

- $\vec{B}$ is the symbol for magnetic field. The units are N/A-m or Tesla.

- The north pole of a bar magnet is the end of the magnet that points in the northern direction if the magnet is suspended, the south pole of the bar magnet is the end of the magnet that points in the southern direction when the magnet is suspended.

- For magnets, like poles repel, unlike poles attract. Note that magnetic field comes out of a north pole, and goes into a south pole.

- The direction of the magnetic field at any point is the same as the direction of the force that acts on the North Pole of a test magnet at the point in question.

- The force that acts on a charged particle moving in a magnetic field is given by $F_m = QvB\sin\theta$, where Q is charge, v is velocity and θ is the angle between the magnetic field vector and the velocity vector.

- The direction of the magnetic force that acts on a charged particle moving in a $\vec{B}$ is found by using the right hand: Extend the fingers of the right hand in the direction of the velocity vector, and allow the fingers to curl through the small angle between $\vec{v}$ and $\vec{B}$ toward $\vec{B}$. The thumb of the right hand then points in the direction of the magnetic force that acts on the charged particle. Note that if charge is negative, the direction of the force is reversed.

- A charged particle moving in a magnetic field executes circular motion. The radius of the motion, r, is found from $r = \dfrac{mv}{QB}$, where Q is the charge of the particle, B is magnitude of magnetic field, m is the mass of the particle, and v is the magnitude of the velocity of the particle.

- When a charged particle is moving in a magnetic field, no work is done; and thus the particle neither gains nor loses energy.

- A velocity selector selects particles according to their velocity. Physically it consists of two parallel metal plates with an electric field $\vec{E} \perp$ to a magnetic field $\vec{B}$, as in the figure. The velocity of the charged particles exiting the velocity selector is given by $v = \dfrac{E}{B}$.

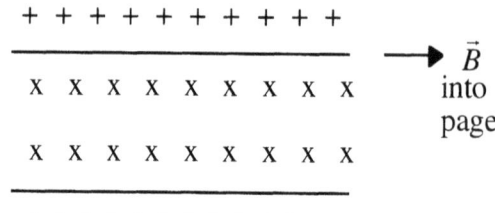

- The force that acts on a current-carrying wire in a magnetic field has its magnitude given by $F = IlB\sin\theta$, where I is current, l is the length of wire in the magnetic field, B is the magnitude of the magnetic field, and θ is the angle between the direction of the current and the magnetic field.

- The direction of the force that acts on a current-carrying wire in a $\vec{B}$ can be found in two ways: Using a right-hand rule: Place fingers of the right hand in the direction of current flow, allow the fingers to curl through the small angle toward $\vec{B}$, the thumb of the right hand then points in the direction of the force.

  The direction of the force that acts on a current-carrying wire is always toward the weakest resultant field. [Note that the wire is in a $\vec{B}$, and the current in the wire produces a $\vec{B}$. These are the two fields one finds where the fields oppose each other.]

- A current produces a magnetic field $\vec{B}$. The direction of the $\vec{B}$ produced by a current I is found using Oersted's Right Hand Rule: If the current carrying wire is grasped with the thumb of the right hand pointing in the direction of conventional current flow, the ends of the fingers point in the direction of the magnetic field produced by the current.

- The magnitude of $\vec{B}$ in the vicinity of a long, straight wire is given by $B = \dfrac{\mu_0 I}{2\pi r}$, where $\mu_0 = 4\pi \times 10^{-7}$ N/A², I is current and r is the distance from the wire.

- The magnitude of the magnetic field at the center of N circular loops of wire is given by
- $B = \dfrac{N\mu_0 I}{2R}$ where N = number of loops, μ_0 is $4\pi \times 10^{-7}$ N/A², I is current, and R is the radius of the circular loops.

- The magnetic field at the center of a solenoid is given by $B = \dfrac{\mu_0 NI}{l} = \mu_0 nI$, where $\mu_0 = 4\pi \times 10^{-7}$ N/A², I is current, N is number of turns of wire and l is the length of the solenoid. Note that n = N/l is a physical parameter called turns per unit length.

SOLVED PROBLEMS

☐☐ **Consider the figure: A long straight wire carries a current I. Find the direction of $\vec{B}$ at point P and point Q.**

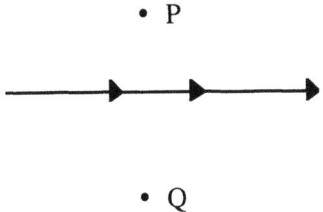

By using Oersted's Right Hand Rule, the direction of $\vec{B}$, at point P is out of the page, and point Q is into the page.

☐☐ **For the previous problem, suppose a proton is moving to the right with velocity v. Find the direction of the magnetic force that acts on the proton if (a) it is at point P and (b) if the proton is at point Q.**

(a) At point P the $\vec{B}$ is out of the page, thus by using the right hand rule, the F_m is down (toward the bottom of the page.)

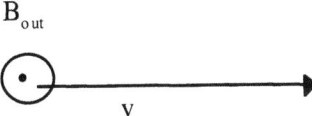

(b) At point Q the $\vec{B}$ is into the page, thus F_m is up (toward the top of the page.)

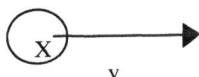

Note that the convention is that if one sees a dot, the $\vec{B}$ or current is coming toward you, if one sees a cross or X, the $\vec{B}$ is going away from you, e.g.,

$\vec{B}$ out of page $\vec{B}$ into page

• • •   X X X

• • •   X X X

◻◻ A proton (mass = 1.67 x 10^{-27} kg) moving at 6 x 10^5 m/s enters a region of $\vec{B}$ in a direction that is perpendicular to the magnetic field. What is the radius of the motion? Assume that b = 0.45 T.

$$r = \frac{mv}{QB} = \frac{(1.67 \times 10^{-27} kg)(6 \times 10^5 m/s)}{(1.6 \times 10^{-19} C)(.45 N/A-m)} = 1.39 \times 10^{-2} m$$

Thus, the radius of the motion is 1.39 x 10^{-2} m or 1.39 cm.

◻◻ A rod that is 1.5 m long is in a region of $\vec{B}$ perpendicular to the magnetic field. If there is a current of 10 A in the rod, and it experiences a force of 1.2 N, find the magnitude of the magnetic field.

F = I(l)Bsin θ ⇒ 1.2 N = 10 A(1.5 m)Bsin 90° ⇒ B = 8 x 10^{-2} N/A-m.
Thus, the magnitude of the magnetic field is 0.08 T.

◻◻ Consider the figure. The long straight wire carries a current I in a region of magnetic field. Find the direction of the force that acts on the wire.

```
    X ↑ X
    X | X
    X ↑ X
    X I X
```

Method 1. Right-hand rule method:

By extending fingers of right hand in the direction of I, and allowing those fingers to curl into $\vec{B}$, the direction of the force that acts is to the left.

Method 2. The direction of the force that acts is always toward the region of least $\vec{B}$. By using Oersted's Right-hand Rule to find the direction of the $\vec{B}$ produced by the current in the wire, the fields oppose on the left side. Hence the force that acts is to the left.

```
    X •  ↑   X X
    X •  |   X X
    X •  |   X X
    X •  ↑   X X
```

☐☐ Find the $\vec{B}$ at point P in the figure.

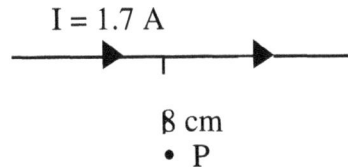

$$B_P = \frac{\mu_0 I}{2\pi r} = \frac{(4\pi \times 10^{-7} N/A^2)(1.7 A)}{(2\pi)(.08m)}$$

$B_P = 4.25 \times 10^{-6}$ N/A-m = 4.25 μT. The direction of $\vec{B}_P$ is into the page.

☐☐ For the previous problem, suppose an electron is moving to the right (in the same direction as the current at point P), with a velocity of 6 x 10⁶ m/s. What is the magnetic force that acts on the electron?

F = QvBsinθ = (1.6 x 10⁻¹⁹ c)(6 x 10⁶ m/s)(4.25 x 10⁻⁶ N/A-m) sin 90°
F = 4.08 x 10⁻¹⁸ N
The direction of the force is down (toward the bottom of the page). A negative charge reverses the direction of the force

☐☐ A flat coil of radius 5 cm has 20 turns. What is the magnitude of the magnetic field at the center of the coil when there is a current of 0.7 A?

$$B_{center} = \frac{N\mu_0 I}{2R} = \frac{(20)(4\pi \times 10^{-7} N/A^2)(.7A)}{2(.05m)}$$

$B_{center} = 1.76 \times 10^{-4}$ N/A-m or 176 μT.

☐☐ Consider the figure. A proton enters the region of $\vec{B}$. Will the proton execute clockwise or counterclockwise motion?

```
       X     X     X

    X     X     X
          →
          v
    ⊕────
    X     X     X
```

From the right hand rule, the force will be such that the particle executes counterclockwise motion. (Note that the centripetal force is provided by the magnetic force.)

☐☐ For the previous problem, suppose the velocity of the proton is 5 x 10⁵ m/s, and the radius of its motion is 24 cm. Find the magnitude of the magnetic field. (The mass of a proton is 1.6 x 10⁻²⁷ kg.)

$$r = \frac{mv}{QB} \Rightarrow B = \frac{mv}{Qr} = \frac{(1.67 \times 10^{-27} kg)(5 \times 10^5 m/s)}{(1.6 \times 10^{-19} c)(.24 m)}$$

B = 2.17 x 10⁻² N/A-m = 21.7 mT

☐☐ In a charge to mass ratio experiment, electrons are accelerated through a potential difference V and then enter a region of magnetic field B. Show that the charge to mass ratio is given by $\frac{Q}{m} = \frac{2V}{B^2 r^2}$, where r is the radius of the motion of the electrons in the magnetic field. Assume that the electrons start from rest.

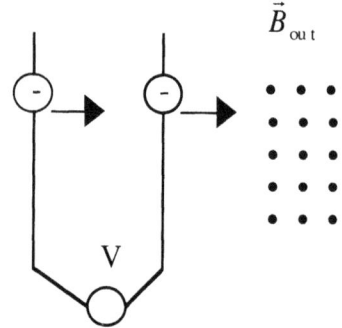

$V = \frac{work}{Q} \Rightarrow work\ done = \Delta energy = \frac{1}{2}mv^2 - 0 = VQ$

Thus, $v^2 = \frac{2VQ}{m}$. Since the motion of charged particles in a magnetic field is circular, the radius of that motion is given by $r = \frac{mv}{Qr}$.

Hence, $r^2 = \frac{m^2 v^2}{Q^2 B^2} = \frac{m^2}{Q^2 B^2}\left(\frac{2VQ}{m}\right) = \frac{m}{Q}\left(\frac{2V}{B^2}\right)$. Thus, $\frac{Q}{m} = \frac{2V}{B^2 r^2}$.

☐☐ Suppose that in a student experiment, the accelerating potential difference was 150 V, the magnitude of the magnetic field was 7.84 x 10⁻⁴ T, and the radius of the motion was 5.4 cm. Find the charge to mass ratio for electrons.

Since in the previous problem, we showed the charge to mass ratio was given by
$\frac{Q}{m} = \frac{2V}{B^2 r^2} = \frac{2(150V)}{(7.84 \times 10^{-4} T)^2 (.054 m)^2} = 1.67 \times 10^{11} c/kg$.

Note that the best value is about 1.7588 x 10¹¹ c/kg.

❏❏ What value does the previous problem give for the mass of the electron, given that the charge of the electron is -1.6×10^{-19} c².

$\frac{Q}{m} = 1.67 \times 10^{11} c/kg \Rightarrow \frac{1.6 \times 10^{-19} c}{1.67 \times 10^{11} c/kg} = m$. Thus, the above experiment gives that the mass of the electron is 9.58×10^{-31} kg.

❏❏ A solenoid has 200 turns and is 5 cm long. Find (a) the parameter turns/length, and (b) the magnitude of the magnetic field at the center of the solenoid if the current in the windings is 0.3 A.

(a) $n = \frac{N}{l} = \frac{200 \, turns}{.05 m} = 4000 \, turns/meter$

(b) $B_{Center} = \frac{\mu_0 NI}{l} = \mu_0 nI = (4\pi \times 10^{-7} N/A^2)(4000 \, turns/m)(0.3 A)$

$B_{center} = 1.51 \times 10^{-3}$ T $= 1.51$ mT

❏❏ Which end of the solenoid in the figure acts as if it were the South Pole of an electromagnet?

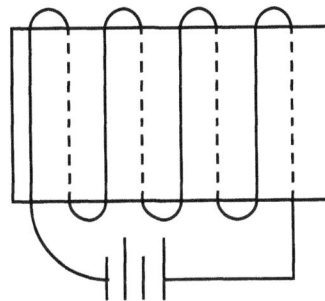

Using Oersted's Right Hand Rule, the direction of $\vec{B}$ within the windings of the solenoid is to the right. Hence the left end of the solenoid in the above figure acts as if it were the South Pole of a magnet.

❏❏ Consider the figure. Will the force that acts on the upper wire be a force of repulsion or a force of attraction?

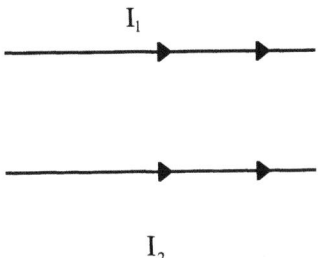

Using either the right hand rule method or the direction of the force is always toward the region of least $\vec{B}$, the force that acts on the upper wire is a force of attraction.

☐☐ Suppose that for the previous problem the upper wire has a length of 10 cm. Let $I_1 = 12$ A, $I_2 = 5$ A, and the distance between the wires is 5 cm. Find the magnitude of the force that acts on the upper wire because it is in a magnetic field created by the current in the lower wire.

$F = I(l)B \sin \theta = I_1 (l) B_2$

$F = I_1(l) \dfrac{\mu_0 I_2}{2\pi r} = \dfrac{(4\pi \times 10^{-7} N/A^2)(.1m)(12A)(5A)}{2\pi(.05m)}$

$F = 2.4 \times 10^{-5}$ N

☐☐ A long, straight wire produces a magnetic field of 48 µT at a distance of 12 cm from the wire. Find the current in the wire.

$B = \dfrac{\mu_0 I}{2\pi r} \Rightarrow I = \dfrac{B(2\pi r)}{\mu_0} = \dfrac{(48 \times 10^{-6} N/A-m)(2\pi)(.12m)}{4\pi \times 10^{-7} N/A^2}$

$I = 28.8$ A

☐☐ Two carbon isotopes ^{12}C and ^{13}C (masses of 19.92 x 10^{-27} kg and 21.59 x 10^{-27} kg, respectively are singly ionized (charge of 1.6 x 10^{-19}c) and exit a velocity selector with a speed of 7 x 10^5 m/s. They then enter the bending region of a mass spectrometer (B = 0.9 T). What is the separation of the two isotopes after they have traveled one half circle?

Here the separation will be the difference in the diameters of the circular motion

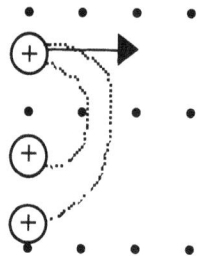

Thus, for ^{12}C: $r_1 = \dfrac{m_1 v}{QB} = \dfrac{(19.92 \times 10^{-27} kg)(7 \times 10^5 m/s)}{(1.6 \times 10^{-19} c)(0.9 N/A-m)}$

$r_1 = 9.68 \times 10^{-2}$ m $\Rightarrow d_1 = 0.194$ m

for ^{13}C: $r_2 = \dfrac{m_2 v}{QB} = \dfrac{(21.59 \times 10^{-27} kg)(7 \times 10^5 m/s)}{(1.6 \times 10^{-19} c)(0.9 N/A-m)}$

$r_2 = 1.05 \times 10^{-1}$ m $\Rightarrow d_2 = 0.201$ m

Thus, the separation is $d_2 - d_1 = 7 \times 10^{-3}$ m = 7 mm

❐❐ A proton at point A in the figure has a speed of 1.4 x 10⁷ m/s. Find the magnitude and direction of the $\vec{B}$ that will cause the proton to move along the semi-circular path as in the figure. (The mass of the proton is 1.67 x 10⁻²⁷ c.)

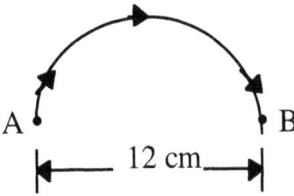

Since the radius of motion for a charged particle moving in a magnetic field is given by r = $\frac{mv}{QB} \Rightarrow B = \frac{mv}{Qr} = \frac{(1.67 \times 10^{-27} kg)(1.4 \times 10^7 m/s)}{(1.6 \times 10^{-19} c)(.06m)}$

Thus, the magnitude of B = 2.43 T. The direction of the magnetic field has to be <u>out</u> of the page for this clockwise motion to occur.

❐❐ (a) Repeat the above problem for an electron travelling from A to B. (The mass of the electron is 9.1 x 10⁻³¹ kg.) (b) How long would it take for the electron to move from A to B?

(a) r = $\frac{mv}{QB} \Rightarrow B = \frac{mv}{Qr} = \frac{(9.1 \times 10^{-31} kg)(1.4 \times 10^7 m/s)}{(1.6 \times 10^{-19} c)(.06m)}$

Thus, the magnitude of B = 1.33 x 10-3 T. The direction of the magnetic field has to be <u>into</u> the page for this clockwise motion to happen.

(b) Disp = vt $\Rightarrow \frac{\pi r}{v} = t = \frac{(3.14)(.06m)}{1.4 \times 10^7 m/s} = 1.35 \times 10^{-8} s$

Thus, it takes 1.35 x 10⁻⁸ s for the e⁻ to travel from A to B.

❐❐ Consider the figure, Find $\vec{B}$ at point P. Both wires are long straight wires with current coming out of the page.

From Oersted's Right Hand Rule, the magnetic field produced by the current in each wire is directed in the down sense at point P (toward the bottom of the page).
Thus, $\vec{B}_{Total} = \vec{B}_1 + \vec{B}_2 \Rightarrow \vec{B}_{Total}$ direction is as discussed above.

$$B_1 = \frac{\mu_0 I_1}{2\pi r_1} = \frac{(4\pi \times 10^{-7} N/A^2)(5A)}{2\pi(.04m)} = 25 \mu T$$

$$B_2 = \frac{\mu_0 I_2}{2\pi r_2} = \frac{(4\pi \times 10^{-7} N/A^2)(10A)}{2\pi(.1m)} = 20 \mu T$$

so that $\vec{B}_P = 45 \mu T$ toward the bottom of the page.

☐☐ **For the previous problem suppose an electron is moving with velocity of 8 x 10⁵ m/s into the page at point P. Find the magnitude and direction of the force acting on the e⁻.**

F = QvBsinθ = (1.6x10⁻¹⁹ c)(8 x 10⁵ m/s)(45 x 10⁻⁶ N/A-m)sin90°
F = 5.76 x 10⁻¹⁸ N. From the right hand rule, the direction of the force is toward the wire carrying current I_1. Hence, $\vec{F} = 5.76 \times 10^{-18}$ N to the right.

☐☐ **Consider the figure. A long, straight wire has a circular loop of radius 8 cm in it. If the current in the wire is 8 A, what is the magnitude of the magnetic field at the center of the loop?**

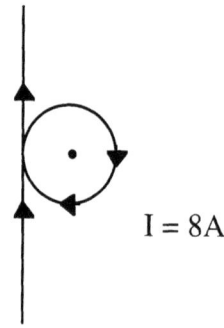

I = 8A

The magnetic field at the center of the loop of wire consists of two parts: (1) the magnetic field due to the long straight wire; and (2) the magnetic field due to the circular loop of wire. Due to Oersted's Right Hand Rule, both magnetic fields are going into the page.
Thus, $B_{center} = B_{straight\ wire} + B_{circular\ loop}$

$$= \frac{\mu_0 I}{2\pi r} + \frac{\mu_0 I}{2R}$$

$$= \frac{(4\pi \times 10^{-7} N/A^2)(8A)}{2\pi(.08m)} + \frac{(4\pi \times 10^{-7} N/A^2)(8A)}{2(.08m)}$$

$B_{center} = 20\ \mu T + 63\ \mu T$
Thus, $\vec{B}$ at the center of the circular loop of wire is 83 μT directed into the page.

❏❏ Which end of the electro-magnet acts as if it were the South Pole of a bar magnet?

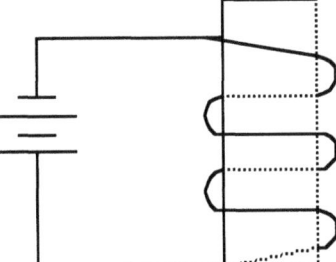

Using Oersted's Right Hand Rule the top end acts as it is were the South Pole of a bar magnet.

❏❏ If the electro-magnet in the figure in the previous problem is to produce a magnetic field at its center of magnitude 15 μT, how much current has to be in the windings? Assume the electro-magnet is 12 cm long.

Since there are three turns,

$$B = \frac{\mu_0 NI}{l} \Rightarrow \frac{Bl}{\mu_0 N} = \frac{(15 \times 10^{-6} N/A-m)(.12m)}{(4\pi \times 10^{-7} N/A^2)(3)} = I$$

I = 0.478 A to produce a 15 μT magnetic field.

❏❏ Consider the figure. Singly ionized Ne atoms (Q = +e) enter the velocity selector as in the figure. Explain why the atoms do not exit this selector.

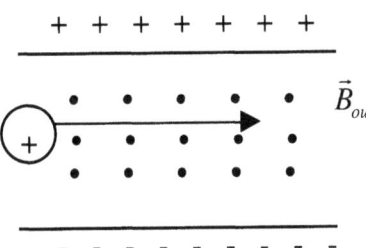

The electric and magnetic forces are both acting in the down sense.

❏❏ If the magnetic field in the previous problem is reversed, the atoms exit with a velocity of 6 x 10⁶ m/s. What must be the magnitude of the electric field if the magnitude of the magnetic field is 0.6 T?

$$v = \frac{E}{B} \Rightarrow E = vB = (6 \times 10^6 \, m/s)(0.6 N/A-m)$$
E = 3.6 x 10⁶ N/c

☐☐ Consider the figure. A wire of length 0.6 m is carrying a current of 1.6 A in a $\vec{B}$ of 0.9 T. Find the force that acts on the wire.

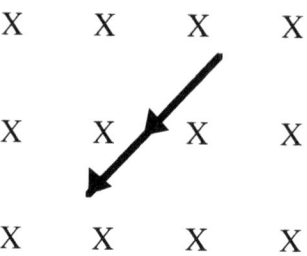

F = I(l)Bsin θ = (1.6 A)(0.6 m)(0.9 T)sin90° = 0.864 N
The direction of the force that acts can be determined to be as indicated by the arrow in the figure by using either the right hand rule method or the direction of the force is toward the region of least magnetic field method.

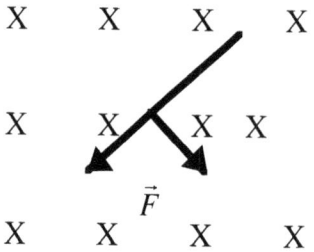

FARADAY'S LAW AND ELECTRO-MAGNETIC INDUCTION

THE FUNDAMENTAL IDEAS ARE THE FOLLOWING:

- Magnetic flux ϕ_B = BAcosθ, where B is magnetic field, A is area and θ is the angle between the area vector $\vec{A}$ and the magnetic field vector $\vec{B}$ (the area vector is $\perp$ to the area and has magnitude equal to the area.)

  Note that for this case ϕ_B = BA (because cos 0° = 1)

  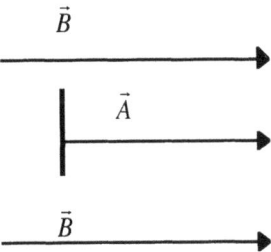

  For this case ϕ_B = 0 (because cos 90° = 0)

  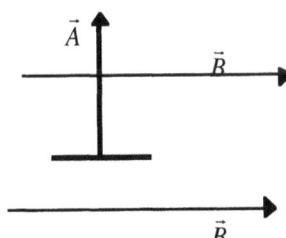

- The unit of magnetic flux ϕ_B is the weber $\left(\dfrac{N-m}{A}\right)$

- Faraday's Law of Electro-magnetic induction says that the average induced emf E is given by $E_{av} = -N\dfrac{\Delta\phi_B}{\Delta t}$, where N is the number of turns, and $\dfrac{\Delta\phi_B}{\Delta t}$ is the change in magnetic flux with respect to time.

- Lenz's Law says that the direction of the induced emf (or current) is such that it opposes by magnetic means the motion that produces it.

- For the case of a rod or wire moving through a magnetic field, the induced emf E is given by E = Blv (B $\perp$ v $\perp$ l), where B is the magnitude of the magnetic field, l is the length of the rod, and v is the velocity of the rod.

Note that this emf is called motional emf, and is produced only if the rod or wire is bending, breaking, or cutting magnetic field lines.

- If for a rod or wire moving through a magnetic field the angle θ between $\vec{v}$ and $\vec{B}$ is not 90°, the induced emf is given by $E = B(l)v\sin\theta$.

- A generator consists of N loops of wire rotating in a magnetic field. The output of the generator is given by $E = NBA\omega\sin\theta = E_{max}\sin(\omega t)$, where N is the number of loops of wire, B is the magnitude of magnetic field, A is the area of the loops of wire, and w is the angular velocity of the rotating loops of wire.

- The output of the generator is sinusoidal and alternating, i.e., the current flows first one way then another.

- For the circuit in the figure, the induced emf E_2 in the #2 circuit is given by $E_2 = M\frac{\Delta i}{\Delta t}$ where M is the coefficient of mutual inductance, and $\frac{\Delta i_1}{\Delta t}$ is the change in current with respect to time in the #1 circuit. The units of M are Henrys $\left(\frac{V}{A/s}\right)$. M depends only on the geometry of the two coils.

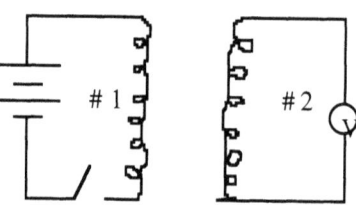

- For the circuit in the figure, the self-induced emf $E = L\frac{\Delta i}{\Delta t}$ where L is the coefficient of self-inductance, and $\frac{\Delta i}{\Delta t}$ is the change in current with respect to time. The units of L are Henrys ($\frac{V}{A/s}$). Note that the single coil is called an inductor with inductance L.

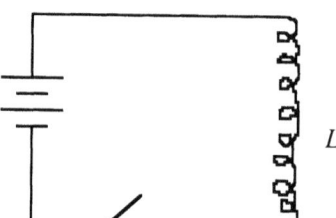

- The energy stored in an inductor is given by $U_L = 1/2 L (I)^2$, where L is the coefficient of self-inductance, and I is the current flowing.

SOLVED PROBLEMS

☐☐ The plane of a loop of wire of area 0.06 m² is parallel to a magnetic field of 0.04 T. Find the magnetic flux ϕ_B that passes through the loop.

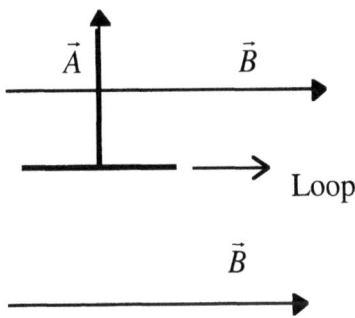

Since $\phi_B = B(A)\cos\theta$

$\phi_B = (0.04\text{T})(0.06\text{m}^2) \cos 90°$

$\phi_B = 0$

☐☐ For the previous problem suppose the loop of wire is rotated clockwise 90°. What is ϕ_B through the coil?

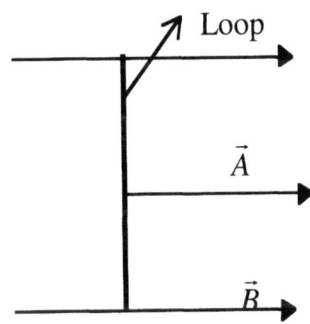

Here, $\phi_B = B(A)\cos\theta$
$= (0.04 \text{ T})(0.06 \text{ m}^2)\cos 0°$
$= 2.4 \times 10^{-3}$ Wb

☐☐ Consider the rod in the figure, which is sliding on frictionless rails. Find the induced emf in the rod if v = 6 m/s l = 0.5 m, and B is 0.04 T.

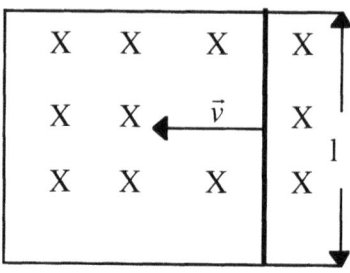

Since v ⊥ l ⊥ B, $E = B(l)v = (0.04 N/A\text{-}m)(6 \text{ m/s})(0.5 \text{ m})$
$E = 0.12 \text{ V}$

☐☐ For the rod in the previous problem, (a) find the direction of the flow of the induced current, and (b) which end of the rod is +?

(a) From Lenz's Law, the direction of the induced current is such that it opposes by magnetic means the motion that produces it, the induced current flows down (toward the bottom of the page).

(b) Since current according to the Franklin convention flows from + → − <u>outside</u> the source, this means that <u>inside</u> the source current flows from − → +. Thus, the lower end of the rod is +.

☐☐ Find the direction of the induced current through the resistor in the figure. The magnet is approaching the stationary coil.

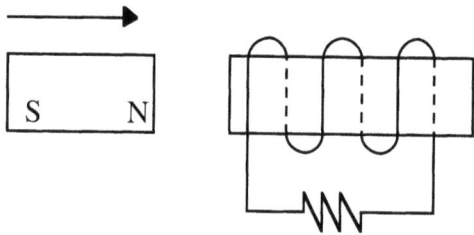

From Lenz's Law, the direction of the induced current is such that it opposes by magnetic means the motion that produced it. Thus, the current produces a magnetic North Pole next to the North Pole of a magnet. This happens (from Oersted's Right-Hand Rule) if the current is going from right to left through the resistor.

▫▫ A circular loop of wire of area 0.7 m² is located in a region of $\vec{B}$ (magnitude 0.8 T) as indicated in the figure. If opposite ends of a diameter are pulled so that the loop of wire is stretched in 0.02 s to where the wire consists of two straight lengths of wire, find the average induced emf.

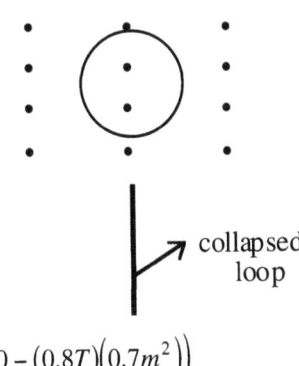

collapsed loop

$$E_{av} = -N\frac{\Delta\phi_B}{\Delta t} = -N\frac{(\phi_B - \phi_{B_0})}{\Delta t} = \frac{-1(0 - (0.8T)(0.7m^2))}{0.02s}$$

$E_{av} = 28$ V

▫▫ For the previous problem, as the wire is collapsing, is the induced current flowing in a clockwise or counter-clockwise manner?

From Lenz's Law, the induced current would flow in a clockwise manner.

▫▫ For the figure, $E_2 = 12$ V when $\frac{\Delta i_1}{\Delta t} = \frac{4A}{s}$. Find the coefficient of mutual inductance between the two coils.

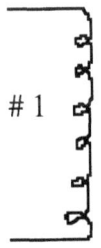

1   # 2

$E_2 = \frac{M\Delta i_1}{\Delta t} \Rightarrow 12V = M\left(\frac{4A}{s}\right)$. Thus, M = $\frac{12V}{4A/s} = 3\frac{V}{A/s} = 3$ Henrys.

❑❑ Consider the circuit in the figure. Find (a) the self-induced emf a long time after the switch was closed, and (b) the energy stored in the coil after a long time.

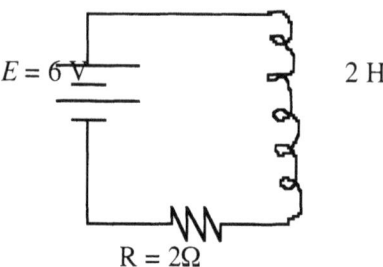

(a) $E = L \frac{\Delta i}{\Delta t}$, after a long time $\frac{(\Delta i)}{\Delta t} = 0$ so the self-induced emf = 0.

(b) After a long time, $I = \frac{V}{R} = \frac{6V}{2\Omega} = 3A$. Thus,

$U_L = (1/2) L I^2 = (1/2) \left(2 \frac{V}{A/s}\right)(3A)^2 = 9J$

❑❑ A metal rod of length 0.8 m moves perpendicularly to a magnetic field of 0.6 T. If an emf of 0.3 V is induced in the rod, how far does the rod move in 6 s, assuming that the velocity of the rod remains constant?

$E = B(l)v \Rightarrow v = \frac{E}{B(l)} = \frac{0.3V}{(0.6T)(0.8m)} = 0.625 m/s$

Thus, x = vt = (.625 m/s)(6 s) = 3.75 m.

❑❑ Consider the figure. What is the direction of the induced current through the resistor if the magnet is moving away from the coil?

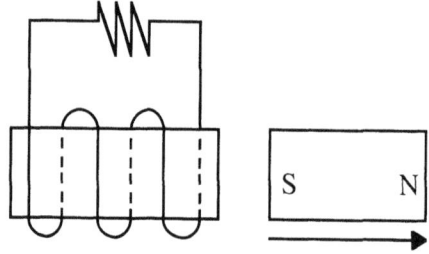

From Lenz's Law, the direction of the induced current has to produce a North Pole on the end of the coil (because this will oppose the motion of the magnet away from the coil). Thus the current will flow from right to left through the resistor.

☐☐ Consider the figure. A rectangular loop of area 0.9 m² is (a) in the xy plane, and (b) in the xz plane. Note that $\vec{B}$ is in the +y sense. Find the magnetic flux for each case.

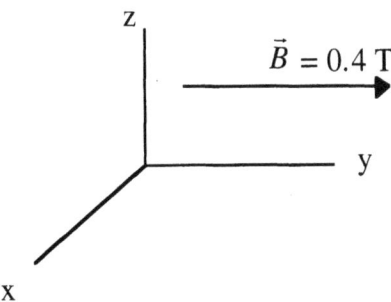

$\phi_B = B(A)\cos\theta$, thus,

(a) Since the loop doesn't enclose any magnetic field if it is the xy plane, $\phi_B = 0$.

(b) Here, $\phi_B = B(A)$
$\phi_B = (0.4 \text{ T})(0.9 \text{ m}^2)$
$\phi_B = 0.36$ Wb

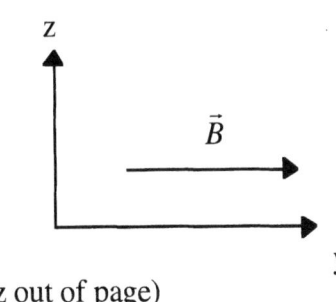

(z out of page)

☐☐ The figure shows a wire, part of which is bent in the shape of a circle. If the radius of the circle is 3.5 cm, what is the average induced emf in the wire if a person pulls on the ends of the wire so that the circle goes to zero in 0.15 s?

$\vec{B}_{out} = 0.8$ T

$E_{av} = -N \dfrac{\Delta \phi_B}{\Delta t}$

$(\phi_B)_0 = B(A) = (0.8 \text{ T})(\pi)(0.035\text{m})^2 = 3.08 \times 10^{-3}$ wb
$(\phi_B) = 0$

Thus, $E_{av} = \dfrac{-1[0 - 3.08 \times 10^{-3} \omega b]}{0.15 S} = 2.05 \times 10^{-2}$ V

❏❏ Show that $\omega b/s = V$

$$\omega b = \left(\frac{N}{A-m}\right)m^2 = \frac{N-m}{A}, \text{ but } A = \frac{C}{s}$$

$$\frac{\omega b}{s} = \frac{N-m}{s\left(\frac{C}{s}\right)} \Rightarrow \frac{\omega b}{s} = \frac{J}{C} = V, \text{ since } 1V = 1\frac{J}{C}$$

❏❏ A generator produces a peak output of 120 V, has 250 turns, and rotates at 377 rad/s. If the rotating coil has an area of 2.25×10^{-2} m², find the magnitude of the magnetic field in which the coil rotates.

For a generator $E = E_{max}\sin\theta = NBA\omega\sin\theta$
Thus, $E = E_{max}$ if $\sin\theta = \pm 1$
Hence, $120 \text{ V} = (250)(B)(2.25 \times 10^{-2} m^2)(377 \text{ rad/s})$
so that $B = 5.66 \times 10^{-2}$ T

❏❏ An inductor has a self-inductance of 0.02 H. If the cross-sectional area of the inductor is 1.4×10^{-3} m², and its length is 0.6 m, how many turns are required? (Note that $\mu_o = 4\pi \times 10^{-7}$ N/A²)

Since for an inductor, $E = L\frac{\Delta I}{\Delta t}$, and Faraday's Law of electro-magnetic induction says that $E = N\frac{\Delta\phi}{\Delta t}$, this implies that $L\frac{\Delta I}{\Delta t} = N\frac{\Delta\phi}{\Delta t}$. Thus, when the switch is closed in the figure:

$$L\frac{(I-0)}{t-0} = \frac{N(\phi-0)}{t-o} \text{ so that } LI = N\phi \Rightarrow N = L\frac{I}{\phi}.$$

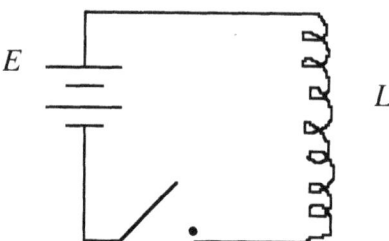

But, $\phi = BA = \left(\frac{\mu_0 NI}{l}\right)A$

so that $N = \frac{LI}{\left(\frac{\mu_0 NI}{l}\right)A} \Rightarrow N^2 = \frac{(L)(l)}{\mu_0 A} = \frac{(0.02H)(0.6m)}{(4\pi \times 10^{-7} N/A^2)(1.4 \times 10^{-3} m^2)}$

$N^2 = 6.82 \times 10^6$

Thus, $N = 2.61 \times 10^3$ turns.

☐☐ A small "search" coil of area 1.8 cm² has 80 turns of fine wire. The coil is placed between the poles of a horseshoe magnet and then suddenly jerked out. If the average induced emf is 0.25 V when the coil is pulled to a field-free region in 0.09 s, what is the magnitude of the magnetic field in the region between the poles of the magnet?

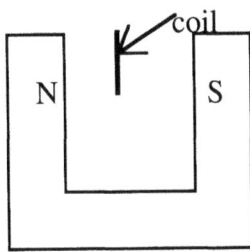

From Faraday's Law of Electromagnetic Induction,

$$E_{av} = -N\frac{\Delta \phi_B}{\Delta t} = -N\frac{[\phi_B - (\phi_B)_0]}{\Delta t}$$

$\phi_B = 0$, $(\phi_B)_0 = BA = B(1.8 \times 10^{-4} m^2)$

thus, $0.25 \text{ V} = \frac{-80[0 - B(1.8 \times 10^{-4} m^2)]}{0.09 s}$ so that B = 1.56 T.

☐☐ An aircraft that has wingspan of 35 m is flying north at 250 m/s, and has an emf of 1.036 V induced between the wing tips. Find the magnitude of the vertical component of the earth's magnetic field at the location of the aircraft.

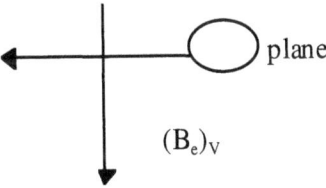

Here, $E = B(l)v \Rightarrow B = \frac{E}{lv} = \frac{1.036V}{(35m)(250m/s)}$

B = 9.87 x 10⁻⁵ T

☐☐ The current in an inductor changes from 25 A to zero in 0.004 s. If the average self-induced emf is 275 V, find (a) the coefficient of self-inductance, and (b) the energy initially stored in the inductor.

(a) $E = L\frac{\Delta i}{\Delta t} \Rightarrow 275 \text{ V} = L\left(\frac{25A - 0}{0.004s - 0}\right)$

L = 4.4 x 10⁻² H

(b) $U_L = (1/2) L I^2 \Rightarrow U_L = (1/2)(4.4 \times 10^{-2} \text{ H})(25 \text{ A})^2$

$U_L = 13.8$ J of energy initially stored.

☐☐ Consider the figure: A rod is moving to the left on resistanceless rails with a speed of 14 m/s. The length of the rod is 0.7 m, and the magnitude of the magnetic field in which the rod is moving is 0.85 T. If the resistance of the rod is 4 Ω, find (a) the current flowing in the rod, and (b) which end of the rod is positive.

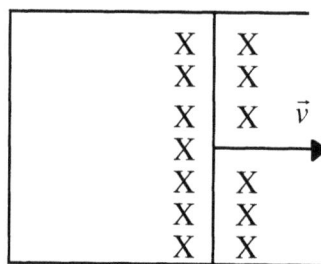

(a) $E = B(l)v = (0.85 \text{ T})(0.7 \text{ m})(14 \text{ m/s}) = 8.33$ V

Since, $I = \dfrac{E}{R} = \dfrac{8.33V}{4\Omega} = 2.08A$

(b) From Lenz's Law, the upper end of the rod is positive.

☐☐ For the circuit in the figure, the coefficient of mutual inductance is 0.4 H. If 24 V is induced in the secondary, what is the time rate of change of current in the primary?

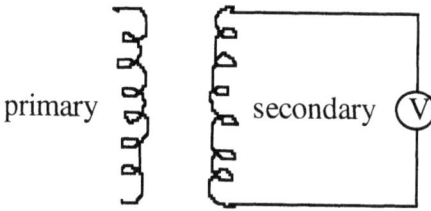

$E_2 = M \dfrac{\Delta i_1}{\Delta t} \Rightarrow 24V = (0.4H)\left(\dfrac{\Delta i_1}{\Delta t}\right)$ so that $\dfrac{\Delta i_1}{\Delta t} = 60 A/s$ is the time rate of change of current in the primary when the induced emf in the secondary is 24 V.

☐☐ A square coil 6 cm on a side has 200 turns and rotates in a magnetic field of 0.15 T. The maximum emf produced is 12 V. Find the angular velocity of the coil.

For a generator E = NBAωsinθ. Thus, for the maximum emf, sinθ = 1 ⇒

$$\omega = \frac{E}{NBA} = \frac{12V}{(200)(0.15N/A-m)(0.06m)^2} = 111/s$$

ω = 111 rad/s

☐☐ For the figure, find the direction of the motion of the magnet if the induced current through the resistor is as indicated.

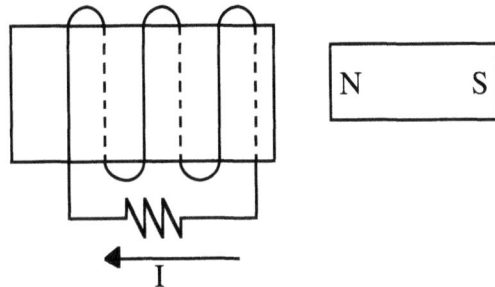

From Oersted's Right-Hand Rule the coil produces a South Pole on the side next to the North Pole of the magnet. From Lenz's Law, the induced current thus opposes the motion producing it, so that the magnet is moving away from the coil.

☐☐ Consider the circuit in the figure. (a) Find the time rate of change of current if the self-induced emf across the coil is 20 V, (b) what will be the current in the circuit after a long time, and (c) what energy will be stored in the coil after a long time?

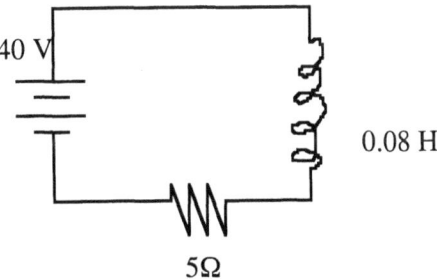

(a) $E = L \frac{\Delta i}{\Delta t} \Rightarrow 20V = (0.08H) \frac{\Delta i}{\Delta t}$. Thus, $\frac{\Delta i}{\Delta t} = 250 A/s$.

(b) After a long time, $\frac{\Delta i}{\Delta t} = 0 \Rightarrow I = \frac{V}{R} = \frac{40V}{5\Omega} = 8A$

(c) $U_L = (1/2)L I^2 = (1/2)(0.08 H)(8 A)^2 = 2.56$ J.

❏❏ A coil of 300 turns and area 350 cm² is placed with its plane perpendicular to the earth's magnetic field, and is rotated in 0.02 s through a fourth turn so that its plane is parallel to the magnetic field of the earth. If the induced emf is 0.36 V, what is the magnetic field of the earth at this location?

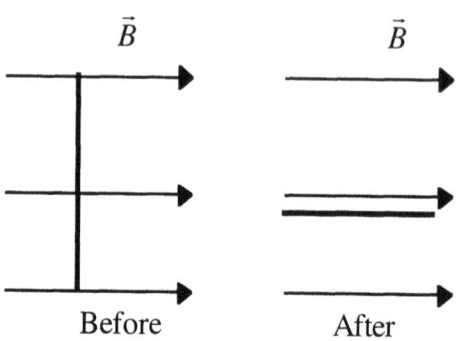

Before   After

$E = -N \dfrac{\Delta \phi_B}{\Delta t} = \dfrac{-N[\phi_B - (\phi_B)_0]}{\Delta t}$; but, $\phi_B = 0$, while $(\phi_B)_0 = BA = B(350 \times 10^{-4} m^2)$.

Thus, $\dfrac{E\Delta t}{NA} = B = \dfrac{(0.36V)(0.02s)}{(300)(350 \times 10^{-4} m^2)} = 6.86 \times 10^{-4} T$

❏❏ The axle of a truck is 2.2 m long. If the truck is moving east with a speed of 20 m/s at a place where the vertical component of the earth's magnetic field is 8×10^{-5} T, find (a) the induced emf between the two ends of the axle, and (b) which end of the truck's axle is negative?

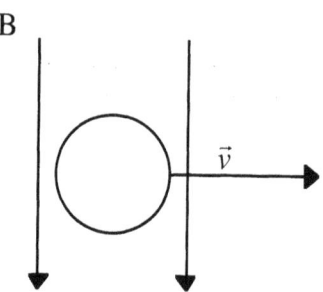

(a) $E = B(l)v = (8 \times 10^{-5} T)(2.2 m)(20 m/s)$
    $E = 3.52 \times 10^{-3}$ V

(b) Since the induced current is flowing into the page, the passenger's side is negative.

☐☐ A coil (see figure) used to create a high voltage has 5,000 turns, and a cross-section of 5 cm². The magnetic field in the core of the coil changes from 1.5 to 0.15 T in 2 x 10⁻⁴ s. Find (a) the original flux through the coil, (b) the final flux through the coil, and (c) the average induced emf in the coil.

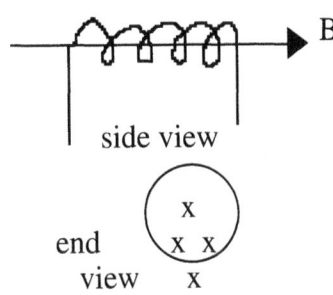

(a) $(\phi_B)_0 = B_0 A$
    $= (1.5 \text{ T})(5 \times 10^{-4} \text{m}^2)$
    $= 7.5 \times 10^{-4}$ Wb

(b) $\phi_B = BA$
    $= (.15 \text{ T})(5 \times 10^{-4} \text{m}^2)$
    $= 0.75 \times 10^{-4}$ Wb

(c) $E_{av} = -N \dfrac{\Delta \phi_B}{\Delta t} = \dfrac{-5000}{2 x 10^{-4} s} \left[0.75 x 10^{-4} - 7.5 x 10^{-4} \right] \omega b$

$E_{av} \approx 16{,}900$ V

ALTERNATING CURRENT AND TRANSFORMERS

THE FUNDAMENTAL IDEAS ARE THE FOLLOWING:

- Direct current (DC) is current that flows in only one direction. DC may be steady state (doesn't change with respect to time) or it may be pulsating.

- Alternating current (AC) is current that flows in first one direction, then in the reverse direction.

- The diagram illustrates AC:

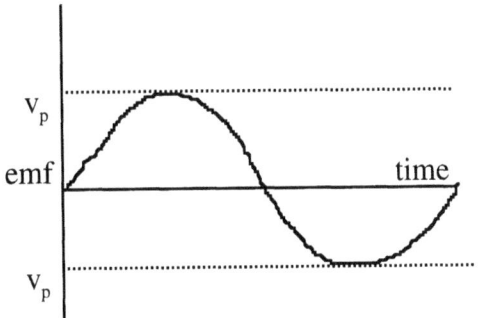

- Some definitions that are associated with AC are:

  1. The peak voltage (v_p) is the maximum potential difference in either the positive or negative sense.

  2. The peak-to-peak voltage (v_{pp}) is the potential difference between peaks. Thus $v_{pp} = 2v_p$.

  3. The instantaneous values of current and potential difference are given by
     $v = v_p \sin\omega t$
     $(i) = i_p \sin\omega t$
     where v_p and i_p are peak values, and ω is angular frequency, while t is time.

  4. The rms (root mean square) value of current and potential difference are given by
     $V = 0.707 v_p$
     $I = 0.707 i_p$
     Note that upper case letters are used for rms values, while lower case letters are used for other values. AC voltmeters and ammeters measure rms values.

- Since frequency is the number of cycles per unit time, AC has associated with it a frequency. In the United States, the frequency of the AC electrical power is 60 Hz.

- Note that from periodic motion, the period (T) is the reciprocal of frequency (f), and that angular frequency (ω) is $2\pi f$. Thus $T = \dfrac{1}{f}$, and $\omega = 2\pi f$.

- A transformer is a device that can be used to step potential difference up or down. Transformers do not work on steady state DC. By applying Faraday's Law of Electromagnetic Induction, one finds that $\frac{E_P}{E_S} = \frac{N_P}{N_S}$ where N_p and N_s are the number of turns on the primary and secondary respectively.

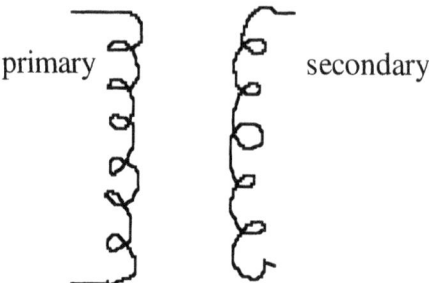

- Since transformers are almost 100 per cent efficient, $P_P \cong P_S \Rightarrow E_P I_P \cong E_S I_S \Rightarrow$ $\frac{E_P}{E_S} = \frac{I_S}{I_P} = \frac{N_P}{N_S}$

- In AC, the AC resistance, or impedance, depends on the circuit component.

- The impedance of a capacitor is given by $X_C = \frac{1}{2\pi f C} = \frac{1}{\omega C}$, where f is frequency, ω is angular frequency, and C is the capacitance in farads. X_C is called capacitive reactance.

- The impedance of an inductor, or coil, is given by $X_L = 2\pi f L = \omega L$, where f is frequency, ω is angular frequency, and L is the self-inductance of the coil. X_L is called inductive reactance.

- The total impedance of a series AC circuit with a resistor, inductor, and capacitor is given by $Z = \sqrt{R^2 + (X_L - X_C)^2}$

- The units of impedance, and inductive and capacitive reactance are ohms.

- If there are no inductors or capacitors in the circuit, Z = R.

- A series circuit is said to be resonant if $X_L = X_C$. The requirement that $X_L = X_C \Rightarrow 2\pi f L = \frac{1}{2\pi f C}$, or the resonant frequency is given by $f = \frac{1}{2\pi\sqrt{LC}} \Rightarrow \omega = \frac{1}{\sqrt{LC}}$

SOLVED PROBLEMS

□□ **In the United States, the power distribution system uses 60 Hz AC. For a potential difference of 120 VAC rms, find (a) the peak potential difference, (b) the peak-to-peak potential difference and (c) the period of the potential difference.**

(a) Since $V_{rms} = 0.707 v_P \Rightarrow v_P = \frac{V_{rms}}{0.707} = \frac{120V}{0.707} = 170V$

(b) $v_{PP} = 2V_P = 340$ V.

(c) $f = 60$ Hz $\Rightarrow T = (1/60)$s $= 1.67 \times 10^{-2}$ s.

❏❏ **A transformer is used to step 120 VAC rms to 6000 VAC. How many turns are on the secondary if the primary winding has 100 turns?**

$$\frac{E_S}{E_P} = \frac{N_S}{N_P} \Rightarrow N_S = \left(\frac{E_S}{E_P}\right) N_P$$

$$N_S = \left(\frac{6000 \text{ VAC}}{120 \text{ VAC}}\right)(100 \text{ turns}) = 5000 \text{ turns}$$

❏❏ **An oscillator consists of an inductor (L = 5 µH) and a capacitor (C = 0.1 µF). Assuming that there is very little resistance, find (a) the resonant frequency and (b) the wavelength at the resonant frequency. (Note that the speed of light, symbol, is 3×10^8 m/s**

(a) $(f) = \dfrac{1}{2\pi \sqrt{LC}} = \dfrac{1}{2\pi\left[(5 \times 10^{-6} H)(0.1 \times 10^{-6} F)\right]^{1/2}}$

$(f) = 2.25 \times 10^5 \dfrac{1}{\left[\left(\dfrac{V}{A/s}\right)\left(\dfrac{C}{V}\right)\right]^{1/2}}$

$= 2.25 \times 10^5$ /s $= 2.25 \times 10^5$ Hz

(b) $c = f\lambda \Rightarrow \lambda = \dfrac{c}{f} = \dfrac{3 \times 10^8 m/s}{2.25 \times 10^5 /s} = 1330 m$

❏❏ **A factory requires 2 MW of power. An AC power plant transmits this power over transmission lines that have a resistance of 2 Ω. How much power is required from the generator if (a) the power is transmitted at 2500 V, and (b) 250,000 V (Assume an ideal transformer).**

(a) $P_{total} = P_{Lines} + P_{factory}$

Since Power $= I^2R \Rightarrow 2 \times 10^6$ watts $= VI$
$= 2 \times 10^6$ watts $= (2500V)I \Rightarrow I = 800$ A

$P_{lines} = I^2R = (800A)^2(2\Omega) = 1.28$ MW,
so that the total power $= 2$ MW $+ 1.28$ MW $= 3.28$ MW.

(b) If the power is transmitted at 250,000 V $\Rightarrow 2 \times 10^6$ watts $= VI$

$\Rightarrow I = \dfrac{2 \times 10^6 \text{ watts}}{250,000 V} = 8A$

$P_{lines} = I^2R = (8A)^2(2 \Omega) = 128$ watts

$P_{total} = 2$ MW $+ 128$ watts ≈ 2 MW

Note : Two reasons that power companies transmit electrical power at a high potential difference is because (1) it decreases power losses in the transmission lines, and (2) the wire that is used to transmit the power is smaller at a high potential difference.

☐☐ Find the reactance of a 54 MH inductor at 60 Hz.

$X_L = 2\pi fL = (6.28)(60 /s)(54 \times 10^{-3} \text{ V/A/s}) = 20.3 \text{ }\Omega$

☐☐ Find the reactance of a 12 µF capacitor at 60 Hz.

$X_C = \dfrac{1}{2\pi fC} = \dfrac{1}{(6.28)(60/s)(12 \times 10^{-6} C/V)}$

$X_C = 221 \text{ }\Omega$

☐☐ At what frequency does a 7 µF capacitor have a reactance of 150 Ω?

$X_C = \dfrac{1}{2\pi fC} \Rightarrow f = \dfrac{1}{2\pi CX_C} = \dfrac{1}{(6.28)(7 \times 10^{-6} F)(150 \Omega)}$

f = 152 Hz

☐☐ Find the capacitive reactance for a 2 µF capacitor at (a) 60 Hz, (b) 6000 Hz, and (c) 6 x 10⁶ Hz.

(a) $X_C = \dfrac{1}{2\pi fC} = \dfrac{1}{(6.28)(60 Hz)(2 \times 10^{-6} F)} \cong 1330 \Omega$

(b) $X_C = \dfrac{1}{2\pi fC} = \dfrac{1}{(6.28)(6000 Hz)(2 \times 10^{-6} F)} \cong 13.3 \Omega$

(c) $X_C = \dfrac{1}{2\pi fC} = \dfrac{1}{(6.28)(6 \times 10^6 Hz)(2 \times 10^{-6} F)} \cong 1.33 \times 10^{-2} \Omega$

Note that X_C decreases as frequency increases.

☐☐ Find the inductive reactance for a 4mH inductor at (a) 60 Hz, (b) 6000 Hz, (c) 6 x 10⁶ Hz.

(a) $X_L = 2\pi fL = (6.28)(60 \text{ Hz})(4 \times 10^{-3} H) = 1.51 \text{ }\Omega$

(b) $X_L = 2\pi fL = (6.28)(6000 \text{ Hz})(4 \times 10^{-3} H) = 151 \text{ }\Omega$

(c) $X_L = 2\pi fL = (6.28)(6 \times 10^6 \text{ Hz})(4 \times 10^{-3} H) = 1.51 \times 10^5 \text{ }\Omega$

Note that X_L increases as frequency increases.

❏❏ A 50 mH coil has an AC current of 0.025 A flowing through it when an AC emf of 2.1 V is applied. What is the frequency of the AC current?

$$V = IX_L \Rightarrow X_L = \frac{V}{I} = \frac{2.1V}{.025A} = 84\Omega$$

Since $X_L = 2\pi fL \Rightarrow f = \frac{X_L}{2\pi L} = \frac{84\Omega}{(6.28)(50 \times 10^{-3} H)}$

$f \cong 268$ Hz

❏❏ A 2 μF capacitor is applied to a 4 V, 60 Hz source. What is the current in the circuit?

$$V = IX_C \Rightarrow I = \frac{V}{X_C} = \frac{V}{\frac{1}{2\pi fC}}$$

Thus, $I = (V) 2\pi fC = (4)(6.28)(60 \text{ Hz})(2 \times 10^{-6} \text{ F})$
$I = 3.01 \times 10^{-3}$ A = 3.01 mA

❏❏ A series RCL circuit has a resistance of 300 Ω, an inductive reactance of 500 Ω, and a capacitive reactance of 400 Ω. Find the impedance Z of this circuit.

$$Z = R^2 + (X_L - X_C^2)$$
$$Z = (300\Omega)^2 + (500\Omega - 400\Omega)^2$$
$$Z \cong 316\Omega$$

❏❏ For the previous problem, the current in the circuit is 0.25 A. What is the emf that is applied to this circuit?

$V = IZ \Rightarrow V = (0.25 \text{ A})(316 \text{ }\Omega) = 79$ V

❏❏ The peak emf of a 400 Hz signal is 15 V. Find (a) v_{PP}, (b) $V_{(rms)}$, and (c) the expression that gives v at any time.

(a) $v_{PP} = 2v_{2P} = 2(15 \text{ V}) = 30$ V

(b) $V_{(rms)} = (0.707)v_P = 10.6$ V

(c) $v = v_P \sin \omega t$
Since $f = 2\pi\omega \Rightarrow \omega = \frac{f}{2\pi} = \frac{400Hz}{6.28} = 63.7/s$
Hence, $v = 15 \sin(63.7 t)$V is the value of the emf at any time t.

❏❏ A 50 Ω resistor (no inductive or capacitive reactance) is connected to a 60 Hz power supply which has a peak emf of 150 V. Find (a) the peak current, (b) the rms current, (c) the power dissipated in the resistor and the expression that gives the value of current at any time.

(a) $i_p = \dfrac{v_P}{R} = \dfrac{150V}{50\Omega} = 3A$

(b) $I(rms) = 0.707\, i_p = 0.707\,(3A) = 2.12\,A$

(c) Power = VI ⇒ (Power)$_{rms}$ = (106 V)(2.12 A)
    (Power rms = 225 watts

Note that the rms value of an alternating current is the value of that direct current which would produce heat at the same rate as the alternating current in a given resistor.

❏❏ Find the capacitive reactance of a 4 μF capacitor at 3000 Hz.

$X_C = \dfrac{1}{2\pi fC} = \dfrac{1}{(6.28)(3000Hz)(4 \times 10^{-6}F)} = 13.3\Omega$

❏❏ For the previous problem, find the inductance required to produce series resonance with the 4 μF capacitor at 3000 Hz.

At resonance $f = \dfrac{1}{2\pi\sqrt{LC}}$ so that $4\pi LCf^2 = 1$

⇒ $L = \dfrac{1}{4\pi^2 Cf^2} = \dfrac{1}{4\pi^2(4 \times 10^{-6}F)(3000Hz)^2}$

$L = 7.04 \times 10^{-4}$ H

❏❏ A radio station has a carrier wave of frequency 1340 kHz. Find the capacitance required to produce resonance at this frequency with a 0.004 H (Assume coil and capacitor are hooked in series).

At resonance, $f = \dfrac{1}{2\pi\sqrt{LC}} \Rightarrow f^2 4\pi^2 LC = 1$

so that $C = \dfrac{1}{f^2 4\pi^2 L}$

$C = \dfrac{1}{(1340 \times 10^3 Hz)^2 (4\pi^2)(0.004H)}$

$C = 3.53 \times 10^{-12}$ F = 3.53 pF

OPTICS
WAVES AND LIGHT

THE FUNDAMENTAL IDEAS ARE THE FOLLOWING:

- A transverse wave is a wave in which the particles that make up the wave vibrate back and forth perpendicularly to the direction of wave travel.

- The wave that spreads out from where a rock hits the water in a pond is an example of a transverse wave.

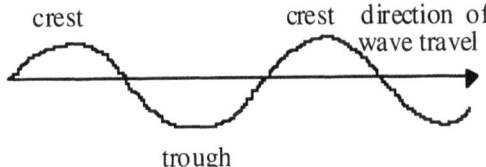

- A longitudinal wave is a wave in which the particles that make up the wave vibrate back and forth along the direction of wave travel.

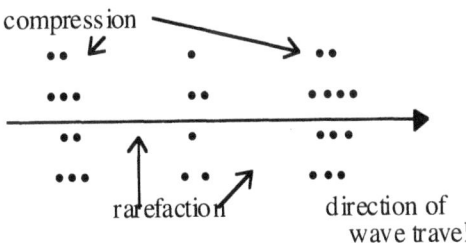

- Sound is an example of a longitudinal wave.

- The distance that a wave travels before it repeats (from crest-to-crest or compression-to-compression) is called wavelength, symbol λ.

- Light acts as if it were a transverse wave.

- Since $x = vt$, $\Rightarrow v = \dfrac{x}{t}$. If $x = \lambda$, then $t = $ period of the wave T. Thus, $v = \dfrac{\lambda}{T} = \lambda f$, since for periodic motion $f = \dfrac{1}{T}$.

- The symbol for the speed of light is c, thus $c = f\lambda$. Experiment has shown that the speed of light in a vacuum (or air) is about 3×10^8 m/s.

- Light is a term that includes the entire electromagnetic spectrum, from radio waves to x-rays. Visible light is a very small part of the electromagnetic spectrum. The human eye can see wavelengths from 400 - 700 nm.

SOLVED PROBLEMS

◻◻ A radio station broadcasts on an assigned frequency of 1340 kHz. What is the wavelength of the wave?

$c = f\lambda \Rightarrow 3 \times 10^8 \text{m/s} = (1340 \times 10^3 \text{ cycles/s}) \lambda$

$\lambda = 224 \text{ m}$

◻◻ Radar works on the principle that the time between sending out a pulse of energy and receiving the reflection (echo) is measured accurately. Find how far away an airplane is if it takes 18 μs for the echo to return to the transmitter.

Since the time it takes for the echo to return is twice the distance that the airplane is from the transmitter, $2x = ct$, so that $2x = (3 \times 10^8 \text{ m/s})(18 \times 10^{-6} \text{ s})$
$\Rightarrow 2x = 5400 \text{ m}$
$x = 2700 \text{ m}$

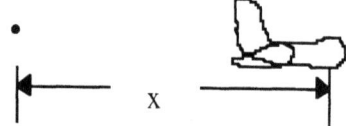

◻◻ A He-Ne laser emits light of wavelength 632.8 nm. What is its frequency?

$c = f\lambda \Rightarrow f = \dfrac{c}{\lambda} = \dfrac{3 \times 10^8 m/s}{632.8 \times 10^{-9} m}$

$f = 4.74 \times 10^{14} \text{ Hz}$.

◻◻ How long would it take a radio signal from Saturn (distance from Earth to Saturn is 1.277 × 10¹² m) to reach Earth? Express your answer in minutes.

$x = vt \Rightarrow 1.277 \times 10^{12} \text{ m} = (3 \times 10^8 \text{ m/s})t$
$t = 4.26 \times 10^3 \text{ s} = 70.9 \text{ minutes}$

◻◻ Red light of λ = 633 nm is incident upon a ruler marked off in mm. How many wavelengths would fit in a distance of 1 mm?

Number of wavelengths $= \dfrac{distance}{\lambda} = \dfrac{1 \times 10^{-3} m}{633 \times 10^{-9} m} = 1580$

Number of wavelengths = 1580.

☐☐ An x-ray machine produces x-rays of frequency 1.4×10^{17} Hz. Find the wavelength of the x-rays.

$c = f\lambda \Rightarrow 3 \times 10^8$ m/s $= (1.4 \times 10^{17}$ Hz$)(\lambda)$
$\lambda = 2.14 \times 10^{-9}$ m $= 2.14$ nm

☐☐ A light year (ly) is the distance that light travels in one year. Find this distance in (a) m, and (b) miles. [Note that 1609 m = 1 mile]

(a) 3×10^8 m/s $\left(\dfrac{86,400 s}{1 day}\right)\left(\dfrac{365 days}{1 year}\right) = 9.46 \times 10^{15}$ m

(b) 9.46×10^{15} m $\left(\dfrac{1 mile}{1609 m}\right) = 5.88 \times 10^{12}$ miles

☐☐ TV Channel 3 broadcasts at a frequency of 63 MHz, while Channel 23 broadcasts at a frequency of 527 MHz. (a) Find the wavelength of each channel and (b) the ratio of the wavelength of Channel 3 to Channel 23.

(a) $c = f\lambda$. Thus, for Channel 3, $\lambda = \dfrac{c}{f} = \dfrac{3 \times 10^8 m/s}{63 \times 10^6 Hz}$

$\lambda_{Channel\ 3} = 4.76$ m

For Channel 23, $\lambda_{Channel\ 23} = \dfrac{3 \times 10^8 m/s}{527 \times 10^6 Hz} = 0.569 m$

(b) $\dfrac{\lambda_{Channel\ 3}}{\lambda_{Channel\ 23}} = \dfrac{4.76 m}{0.569 m} = 8.37$

☐☐ Albert Michelson measured the speed of light by using the apparatus in the figure. Light travels from the rotating mirror to the stationary mirror and back again. While the light is travelling from the rotating mirror to the stationary mirror and back again, the rotating mirror rotates 1/8 revolution. If the distance between the two mirrors is 35.4 km, how fast in rev/min must the rotating mirror rotate?

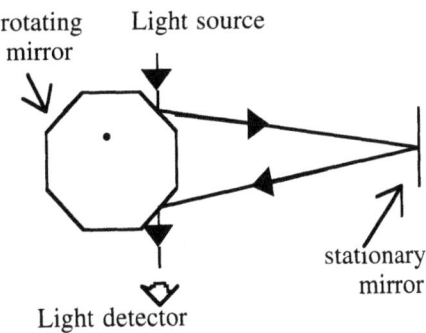

Since $x = vt$, this implies that while the rotating mirror is rotating 1/8 revolution, the light is travelling 70.8 km. Thus $t_{1/8\ rev} = t_{70.8\ km}$. Since $x = vt$,

70.8 x 10³ m = 3 x 10⁸ m/s(t)

t = 2.36 x 10⁻⁴ s for 1/8 rev

t = 2.36 x 10⁻⁴ s for light to travel 70.8 km.

Therefore, 1/8 rev = 2.36 x 10⁻⁴ s ⇒ 1 rev = 1.89 x 10⁻³ s ⇒ that the rotating mirror is rotating at 530 rev/s, so that finally, the rotating mirror is rotating at (530 rev/s)(60 s/1 min) 31,800 rev/min.

⬜⬜ **Suppose that in the previous problem the rotating mirror was 12 sided rather than 8 sided. What would then be the revolutions per minute of the rotating mirror?**

$t_{1/12 \, revz} = t_{70.8 \, km}$. Since x = vt, 70.8 x 10³ m = (3 x 10⁸ m(t)

$$t = 2.36 \times 10^{-4} \text{ s for } 1/12 \text{ rev}$$

$$t = 2.36 \times 10^{-4} \text{ s for light to travel 70.8 km}$$

Therefore 1/12 rev = 2.36 x 10⁻⁴ s ⇒ 1 rev = 2.83 x 10⁻³ s. Thus, the mirror is rotating at 353 rev/s. Finally, (353 rev/s)(60 s/1 min) = 21,200 rev/min.

⬜⬜ **A color of light has 20,000 waves per centimeter. Find (a) the wavelength of the light and (b) its frequency.**

(a) # of waves = $\dfrac{dis\tan ce}{\lambda}$ ⇒ $20,000 = \dfrac{1 \, cm}{\lambda}$

λ = 5 x 10⁻⁵ cm or 500 nm

(b) c = fλ ⇒ $f = \dfrac{3 \times 10^8 \, m/s}{5 \times 10^{-7} \, m}$

f = 6 x 10¹⁴ Hz

⬜⬜ **The FM radio band is assigned to the 88 - 108 MHz frequency band. Find the range of wavelengths that correspond to the FM radio band.**

c = fλ ⇒ $\lambda = \dfrac{c}{f}$

for 88 MHz: $\lambda = \dfrac{3 \times 10^8 \, m/s}{88 \times 10^6 \, Hz} = 3.41 m$

for 108 MHz: $\lambda = \dfrac{3 \times 10^8 \, m/s}{108 \times 10^6 \, Hz} = 2.78 m$

Thus, the range of wavelengths that correspond to the FM radio band is from 2.78 to 3.41 m.

▫▫ A folded dipole is an antenna that is used to maximize signal reception. (See figure) It is used to receive a signal from a 90 MHz transmitter. Find how long it should be.

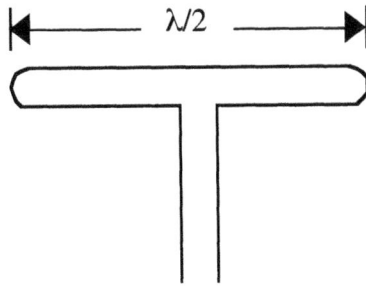

$$c = f\lambda \Rightarrow \lambda = \frac{c}{f} = \frac{3 \times 10^8 \, m/s}{90 \times 10^6 \, /s} = 3.33 m$$

Thus, $\lambda/2 \cong 1.67$ m

GEOMETRIC OPTICS I

THE FUNDAMENTAL IDEAS ARE THE FOLLOWING:

- A ray of light is a line drawn in the direction of wave travel

- For reflection, the incident ray makes an angle relative to the normal called the surface angle of incidence, and the reflected ray makes an angle with respect to the normal called the angle of reflection.

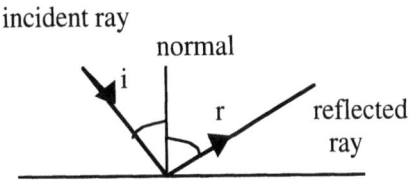

reflecting surface

- Refraction is the bending of light as it travels from one transparent medium to another. Again, the angles are measured relative to the normal, and here the angle of refraction is the angle that the refracted ray makes with respect to the normal.

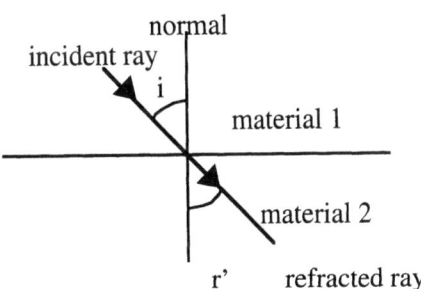

- There are three Laws of Geometric Optics:

  I. The incident ray, the reflected ray, the refracted ray, and the normal to the surface all lie in the same plane.

  II. For reflection, the angle of incidence equals the angle of reflection.

  III. (Snell's Law) For light coming from a vacuum or air entering a material,
  $$n = \frac{\sin i}{\sin r} = \frac{c}{V} = \frac{\lambda_{vacuum\,or\,air}}{\lambda_{material}}$$, where V is speed of light in a material, and n is called the index of refraction of the material.

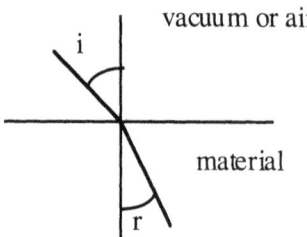

Note that Snell's Law can be rewritten for light coming from material A and going to material B as $\frac{n_B}{n_A} = \frac{\sin\theta_A}{\sin\theta_B} = \frac{V_A}{V_B} = \frac{\lambda_A}{\lambda_B}$. This gives a general form of Snell's Law;

$n_A \sin\theta_A = n_B \sin\theta_B$

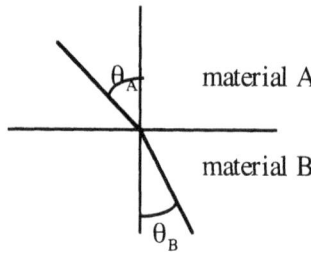

- When light enters a material that has a higher index of refraction, it is refracted toward the normal.

- When light enters a material that has a lower index of refraction, it is refracted away from the normal.

- The index of refraction of a material when viewed almost normally from a point in air is given by n = (actual depth)/(apparent depth).

- The critical angle is that angle of incidence for which the angle of refraction is 90°.

- If the angle of incidence is greater than the critical angle, total internal reflection occurs.

- The function of any optical instrument is to form an image.

- To describe the character of the image means to say whether the image is real or virtual, erect or inverted, and magnified or reduced in size.

- A virtual image cannot be projected on a screen, it appears to come from or to be at a particular place.

- Observations about the image formed by a flat surface plane mirror are:
  a. The image is the same size as the object.
  b. The image is virtual.
  c. The image is erect.
  d. The image is just as far behind the mirror as the object is in front of the mirror.
  e. There is reversal with respect to the reflecting plane - if the person in front of the mirror raises her right hand, the image raises its left hand.

- A concave mirror is a portion of a spherical surface that has had the " inside" polished.

- For the concave mirror in the figure,
    cV is the principal axis
    FV is the focal length of the mirror, f
    c is the center of curvature of the mirror
    R is the radius of curvature of the mirror
    F is the principal focus of the mirror

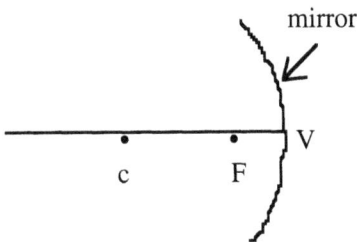

- The distance the object is from the mirror is the object distance p.

- The distance the image is from the mirror is the image distance q.

- The mathematical relationships $\frac{1}{p} + \frac{1}{q} = \frac{1}{f}$ and

    linear magnification = $\frac{A \text{ linear dimension of the image}}{A \text{ corresponding linear dimension of the object}}$, OR

    l.m. = $\frac{q}{p} = \frac{\text{height of image}}{\text{height of object}}$, find the location and size of the image.

- Note that $f \approx R/2$

- A convex mirror is a portion of a spherical surface where the "outside" has been polished.

- For the convex mirror in th figure,
    cV is the principal axis.
    FV is the focal length of the mirror, f.
    c is the center of curvature of the mirror.
    R is the radius of curvature of the mirror.
    F is the principal focus of the mirror.

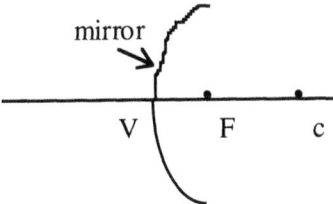

- The mathematical relationships $\frac{1}{p} + \frac{1}{q} = \frac{1}{f}$ and l.m. = $\frac{q}{p} = \frac{\text{ht of image}}{\text{ht of object}}$ find the location and size of the image.

- $f \approx R/2$.

- The standard rays for mirrors are the following (convex case in parentheses):
    a. The ray parallel to the principal axis is reflected so that it passes through (or appears to come from) the principal focus of the mirror.

b. The ray that passes through (or would pass through) the principal focus is reflected back parallel to the principal axis.

c. The ray that passes through (or would pass through) the center of curvature is reflected back upon itself.

- The sign convention for mirrors is:
  a. The focal length of a convex mirror is negative.

  b. The image distance is negative if the rays of light go away from the image after leaving the mirror.

SOLVED PROBLEMS

❑❑ **A ray of light in air is incident at an angle of incidence of 45° upon a piece of glass (n = 1.6). Find the angle of refraction.**

$n = \dfrac{\sin i}{\sin r}$ for light coming from air or a vacuum into a material. Thus,

$\sin r = \dfrac{\sin i}{n} = \dfrac{\sin 45°}{1.6}$. Hence, $\sin r = 0.442 \Rightarrow r = 26.2°$. The angle of refraction is 26.2°.

❑❑ **When light enters a material, part of it is reflected and part of it refracted as in the figure. If the angle of incidence is 60°, what must be the index of refraction of the material so that the angle between the refracted and reflected ray is 90°?**

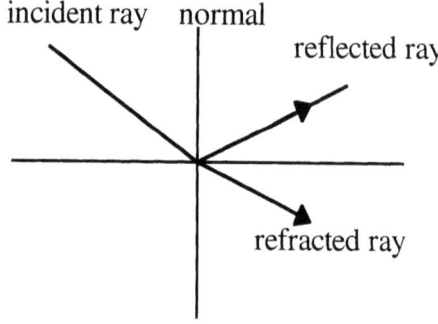

From the fact that the angle of incidence equals the angle of reflection, and the fact that the angle between the reflected and refracted ray is 90°, the angle of refraction is 30° (a straight line has an angle of 180°). Thus, $n = \dfrac{\sin i}{\sin r} = \dfrac{\sin 60°}{\sin 30°} = 1.73$

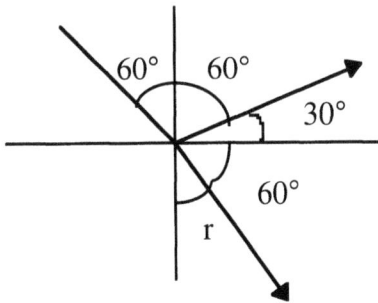

☐☐ A piece of glass has an index of refraction of n = 1.5. What should be the angle of incidence i such that the angle of refraction is i/2?

$n = \dfrac{\sin i}{\sin r} \Rightarrow n \sin r = \sin i \Rightarrow 1.5 \sin\left(\dfrac{i}{2}\right) = \sin i$

from trigonometry $\sin 2\theta = 2 \sin\theta \cos\theta$

Thus, $1.5\sin(i/2) = 2\sin(i/2)\cos(i/2) \Rightarrow \dfrac{1.5}{2} = \cos\left(\dfrac{i}{2}\right) \Rightarrow \cos\left(\dfrac{i}{2}\right) = .75$, so that $i/2 = 41.4°$. Finally $i = 82.8°$.

☐☐ Show that when a plane mirror is rotated through an angle θ about an axis perpendicular to the plane of incidence a ray of light reflected from the mirror is rotated through an angle 2θ.

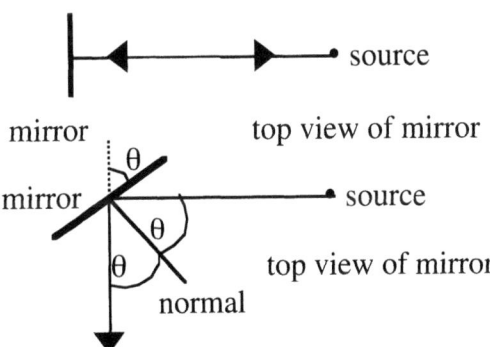

Hence, the light is reflected through an angle 2θ because the angle of incidence is θ, the angle of reflection is θ, and the light is reflected through an angle 2θ.

☐☐ Light comes from cs_2 (n = 1.63) and enters water (n = 1.33). If the angle of incidence is 45°, what is the angle of refraction?

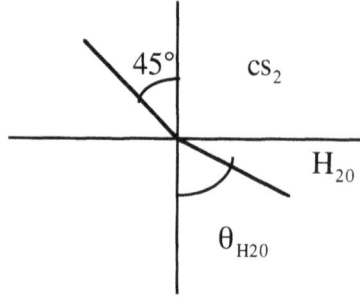

Using Snell's Law in its most general form $n_A \sin\theta_A = n_B \sin\theta_B$
$(1.63)\sin 45° = 1.33 \sin\theta_{H20}$
$0.866 = \sin\theta_{H20}$
$\theta_{H20} \cong 60°$
so that the angle of refraction is 60°.

☐☐ What is the critical angle for a water-air interface ($n_{H20} = 1.33$, $n_{air} \cong 1$)?

The critical angle of incidence has an angle of refraction of 90°. Thus,
$n_A \sin\theta_A = n_B \sin\theta_B$
$1.33 \sin\theta_c = (1)\sin 90°$
$\sin\theta_c = 1/1.33 = 0.75$
Hence, $\theta_c = 48.6°$; the critical angle of incidence for a water-air interface is 48.6°.

☐☐ By drawing the standard rays for the mirror in the figure, describe the character of the image.

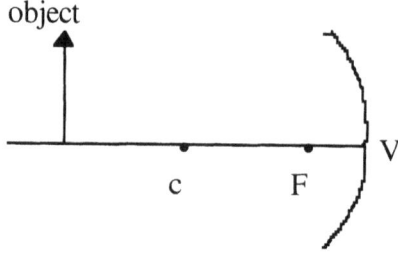

The image is inverted, real and reduced in size.

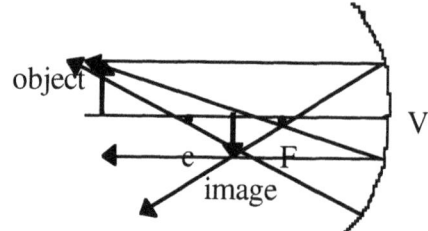

❑❑ **Repeat the above question for the mirror in the figure:**

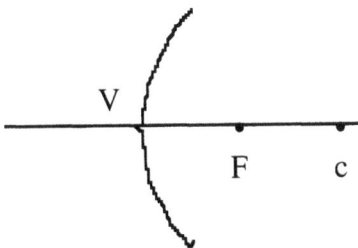

The image is erect, reduced, and virtual.

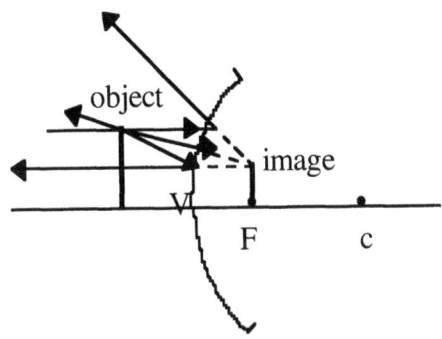

❑❑ **A magnifying mirror of focal length 10 cm produces an erect, virtual image that is magnified by a factor of 3. Find (a) the object distance, and (b) the image distance.**

(a) Here, the object is inside the principal focus of the mirror, and image distance is -.

Thus, l.m. = $\frac{q}{p} = 3x \Rightarrow q = -3p$

Hence, $\frac{1}{f} = \frac{1}{p} + \frac{1}{q} \Rightarrow \frac{1}{10cm} = \frac{1}{p} - \frac{1}{3p}$, so that $\frac{1}{10cm} = \frac{2}{3p} \Rightarrow p = 6.67cm$

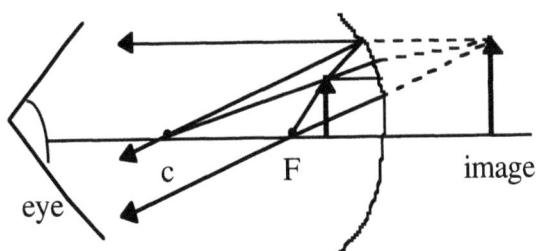

(b) Since $q = -3p \Rightarrow q = -20$ cm

□□ A fortune teller's eye is 30 cm from the center of a polished glass sphere of radius 10 cm. (a) Where is the image of her eye located, and (b) what is the linear magnification of this mirror?

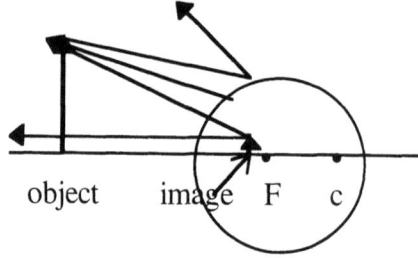

(a) Here, $f = R/2 = 5$ cm, but, the focal length of a convex mirror is $-$ $\Rightarrow$ $f = -5$ cm and $p = 25$ cm. Thus, $\frac{1}{p} + \frac{1}{q} = \frac{1}{f} \Rightarrow \frac{1}{q} = \frac{1}{f} - \frac{1}{p} = -\frac{1}{5} - \frac{1}{25}$

$q = -4.17$ cm

Note that the sign convention says that if the rays go away from the image after leaving the lens, the image distance is negative. Thus, the math and the sign convention agree.

(b) $.m. = \frac{q}{p} = \frac{-4.17 cm}{25 cm} = 0.167 X$

□□ An object in a swimming pool filled with water (n = 1.33) appears to be 1.4 m below the surface. What is the actual depth?

$n = \frac{actual\ depth}{apparent\ depth}$, so that actual depth $= n\ (\ apparent\ depth)$

$= (1.33)(1.4\ m)$

Hence, the actual depth is 1.86 m.

□□ A clinic that treats anorexia installs a mirror that will make a patient appear to be 1.1 times larger than they actually are. If the patient views herself at a distance of 1.9 m from the mirror, (a) what kind of mirror is needed, (b) find the focal length of the mirror, and (c) find the radius of curvature of the mirror.

(a) A <u>concave</u> mirror will form an erect, magnified, virtual image when the object is inside the principal focus. This can be seen by drawing a ray diagram for this case.

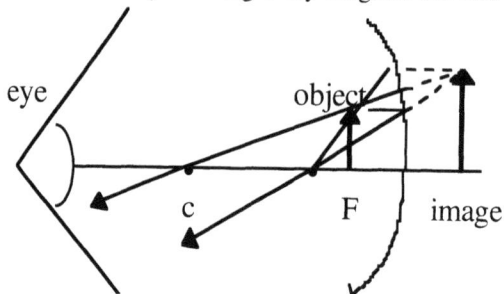

(b) Here l.m. = $\frac{q}{p}$ = 1.1X $\Rightarrow$ q = -1.1p = -2.09 m. The minus for q comes from the sign convention, which says that if the rays go away from the image when they leave the mirror, q is negative.

Hence, $\frac{1}{f} = \frac{1}{p} + \frac{1}{q} \Rightarrow \frac{1}{f} = \frac{1}{1.9m} - \frac{1}{2.09m}$

f = 20.9 m

(c) R = 2f $\Rightarrow$ R = 41.8 m

☐☐ **A clinic that treats overweight patients installs a mirror that makes a patient look 95% as large as they are. If the patient is to stand 1.8 m from the mirror, (a) what kind of mirror is needed, (b) find the focal length of the mirror, and (c) find the radius of curvature of the mirror.**

(a) A <u>convex</u> mirror will give an image that is erect, virtual, and reduced in size, as can be seen from the sketch.

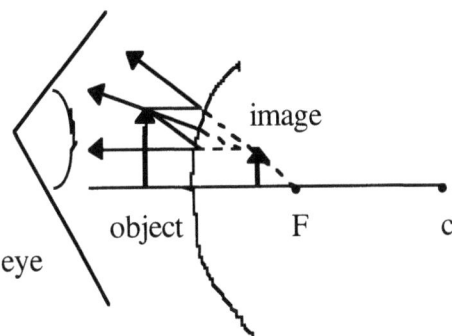

(b) Here, l.m. = $\frac{q}{p}$ = 0.95 $\Rightarrow$ q = -0.95p

q = -1.71 m

Again, the image distance is negative because the rays go away from the image when they leave the mirror. Thus, $\frac{1}{f} = \frac{1}{p} + \frac{1}{q} = \frac{1}{1.8m} - \frac{1}{1.71m}$

f = -34.2 m.

Note that the math and the sign convention agree on the sign of the focal length.

(c) R = 2f $\Rightarrow$ R = - 68.4 m or the radius of curvature of the mirror is 68.4 m.

☐☐ **Blue Light in air has a wave length of 410 nm. Find its wave length in glass (n = 1.6)**

n = $\frac{\lambda_{air}}{\lambda_{material}}$ for light coming from air going to a material.

Thus, $\lambda_{material} = \frac{\lambda_{air}}{n} = \frac{410nm}{1.6}$ = 256nm

☐☐ **Consider the figure. A flat glass slab of thickness t has as ray of light fall on the slab with an angle of incidence i and an angle of refraction r. (a) Show that the angle of refraction at the second interface is i, and (b) show that the lateral displacement d of the ray is given by**
$$d = \frac{t\sin(i-r)}{\cos r}.$$

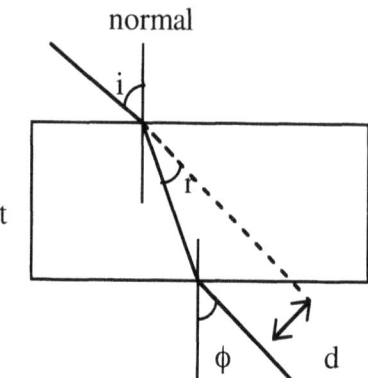

(a) At the first interface, $n = \frac{\sin i}{\sin r} \Rightarrow \sin r = \frac{\sin i}{n}$. The angle of incidence at the 2nd interface is r. Thus, $n_g \sin r = (1)\sin\phi$. But from the 1st interface, n sinr = sin(i). Thus $\phi = i$, and the light exits the 2nd interface with an angle of refraction i.

(b) Consider the figure: (Δ ADC)

sin(i-r) = DC/AC

Thus, DC = d = AC sin(i-r).

Looking at ΔABC, cos r = AB/AC = t/(AC) ⇒ AC = t/(cos r).

Hence, $d = \frac{t}{\cos r} \sin(i-r)$

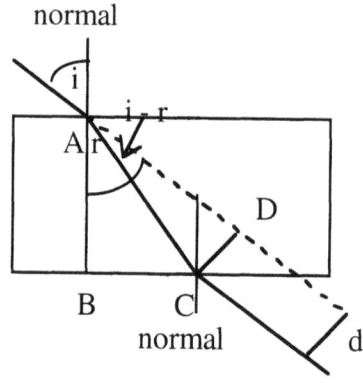

☐☐ **A 6 cm tall object is 12 cm from a convex mirror of focal length 4 cm. How tall is the image?**

Here f is negative because of the sign convention.
$$\frac{1}{f} = \frac{1}{p} + \frac{1}{q} \Rightarrow \frac{1}{-4cm} = \frac{1}{12cm} + \frac{1}{q}$$

$q = -3$ cm $\Rightarrow$ l.m.$= \dfrac{q}{p} = \dfrac{-3cm}{12cm} = \dfrac{1}{4}x$

Since l.m. $= \dfrac{ht\ of\ I}{ht\ of\ O} \Rightarrow ht\ of\ I = \left(\dfrac{1}{4}\right) 6cm$, so that the height of the image is 1.5 cm.

☐☐ **Repeat the previous problem if the mirror is concave.**

Here f = 4 cm, p = 12 cm.

$\dfrac{1}{p} + \dfrac{1}{q} = \dfrac{1}{f} \Rightarrow \dfrac{1}{q} = \dfrac{1}{f} - \dfrac{1}{p} = \dfrac{1}{4cm} - \dfrac{1}{12cm}$, so that q = 6 cm.

l.m. $= \dfrac{q}{p} = \dfrac{6cm}{12cm} = \dfrac{1}{2}X$. Thus, the image is 3 cm tall.

☐☐ **Show for an eye looking almost vertically in a material (e.g. water) and an object in air, that the index of refraction of the material n is given by n = $\dfrac{apparent\ height}{actual\ height}$**

Consider the figure.

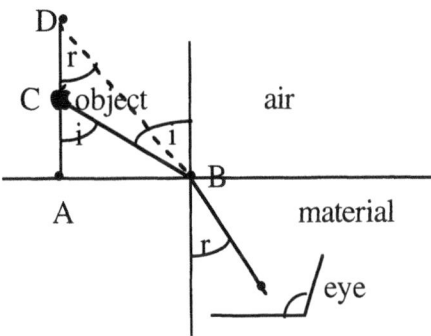

From Snell's Law, $n_{air} \sin(i) = n_{mat} \sin(r)$. Thus, $n_{mat} = \dfrac{\sin i}{\sin r} = \dfrac{AB/CB}{AB/DB}$.

$n_{mat} = \dfrac{DB}{CB} \approx \dfrac{DA}{CA}$ if looking vertically. Thus, $n_{mat} = \dfrac{apparent\ height}{actual\ height}$. Since DA is apparent height and CA is actual height.

☐☐ **A diver is 3 m below the surface of a lake. A bird flies overhead 4 m above the surface of the lake. When the bird is overhead, how far above the diver does it appear to be?**

$n_{H20} = \dfrac{apparent\ height}{actual\ height} \Rightarrow$ apparent height = (n_{H20}) actual height

$\phantom{n_{H20} = \dfrac{apparent\ height}{actual\ height} \Rightarrow} = (1.33)4m$

$\phantom{n_{H20} = \dfrac{apparent\ height}{actual\ height} \Rightarrow} = 5.32$ m above the surface of the lake.

Hence, the bird will appear to be 5.32 m + 3 m = 8.32 m above the diver.

🔲 **For which position will an object in front of a convex mirror be (a) real and (b) virtual?**

The image of a convex mirror is always virtual, reduced, and erect. This can be seen by sketching the ray diagrams for different positions. Thus,

(a) No position
(b) All positions

🔲 **Repeat the above problem for a concave mirror.**

The image of a concave mirror will be real for the object distance greater than the focal length. For object distances less than the focal length, the image will be virtual. If the object distance equals the focal length, the image is formed at infinity. Again, by drawing a few ray diagrams, this will be apparent.

🔲 **An object 4 cm tall is 10 cm from the vertex of a convex mirror whose radius of curvature is 12 cm. (a) Sketch the ray diagram and from the sketch describe the character of the image, and (b) determine the height of the image.**

(a) From the sketch, the image is virtual, reduced and erect.

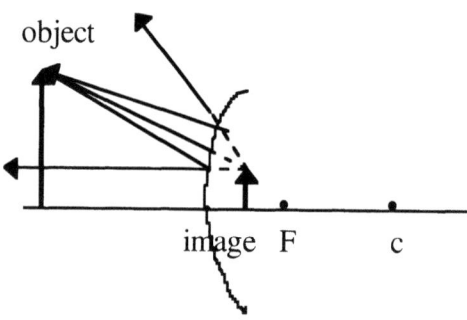

(b) l.m. $= \dfrac{q}{p}$. Thus, $\dfrac{1}{f} = \dfrac{1}{q} + \dfrac{1}{p} \Rightarrow \dfrac{1}{q} = \dfrac{1}{f} - \dfrac{1}{p}$. Hence, $\dfrac{1}{q} = \dfrac{-1}{6cm} - \dfrac{1}{10cm}$.

Note that f = R/2 = − 6 cm
q = −3.75 cm

Thus, l.m. $= \dfrac{q}{p} = \dfrac{-3.75cm}{10cm} = 0.375X = \dfrac{ht\ of\ I}{ht\ of\ object}$

so that ht of I = (0.375)(ht of object)
= (0.375)(4 cm)
= 1.5 cm
The height of the image is 1.5 cm.

☐☐ **Repeat the above question if the mirror is concave.**

(a) f = R/2 = 6 cm

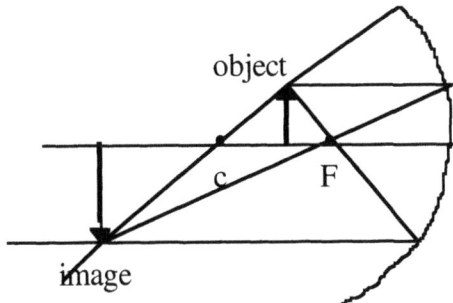

From the sketch the image is real, inverted, and magnified.

(b) l.m. = $\frac{q}{p}$. Thus, $\frac{1}{f} = \frac{1}{q} + \frac{1}{p} \Rightarrow \frac{1}{q} = \frac{1}{f} - \frac{1}{p}$ so that

$\frac{1}{q} = \frac{1}{6cm} - \frac{1}{10cm} \Rightarrow q = 15cm$

l.m. = $\frac{q}{p} = \frac{15cm}{10cm} = 1.5X$ so that (ht of I) = (l.m.)(ht of object)

= (1.5)(4 cm) = 6 cm

☐☐ **A ray of light is incident upon the interface separating glass (n = 1.6) and water (n = 1.33). Find the angle of refraction if the ray has an angle of incidence of 30° and (a) originates in the glass, and (b) originates in the water.**

(a) Using Snell's Law, $n_g \sin\theta_g = n_{H20} \sin\theta_{H20}$
 (1.6)sin30° = (1.33)sinθ_{H20} ⇒ θ_{H20} = 37°

(b) $n_g\sin\theta_g = n_{H20} \sin\theta_{H20}$
 ⇒ (1.6)sinθ_g = (1.33)sin30°
 θ_g = 24.6°

☐☐ **For the previous question, (a) from which material must the light come in order for total internal reflection to take place, and (b) what is the critical angle for this combination?**

(a) For total internal reflection to take place, the light has to enter the material with the lower index of refraction.

(b) $n_g\sin\theta_g = n_{H20}\sin\theta_{H20}$
 (1.6)sinθ_c = (1.33)sin90°
 Thus, sinθ_c = (1.33)/(1.6) = 0.831
 θ_c = 56.2° is the critical angle of incidence for the glass water interface.

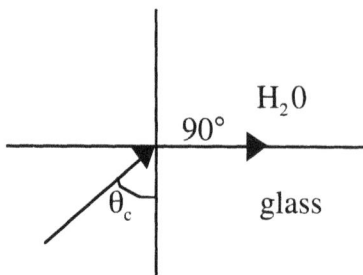

☐☐ **A point source of light is 30 cm below the surface of a pond. Find the diameter of the largest circle at the surface through which light can emerge from the water.**

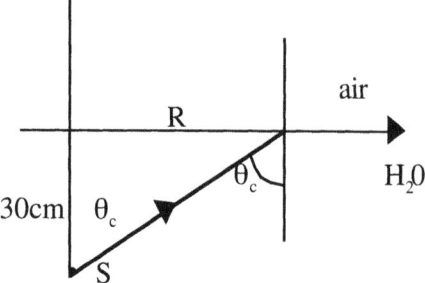

By using Snell's Law and the definition of the critical angle of incidence, the critical angle for a water-air interface is 48.6°. Thus, from the figure, the light will not escape from the water if the angle of incidence is greater than the critical angle. Hence,

$\tan\theta_c = \dfrac{R}{30cm} \Rightarrow R = (\tan 48.6°)(30cm) = 34cm$, so that the diameter of the circle of light (or largest circle from which light can escape) is 68 cm.

GEOMETRIC OPTICS II

THE FUNDAMENTAL IDEAS ARE THE FOLLOWING:

- A spherical lens is a transparent material whose faces are portions of a spherical surface.

- A lens works on the basis of refraction.

- A converging lens:
  i. Brings parallel light rays together.
  ii. Is thicker in the middle.
  iii. Is called a convex lens.

- A diverging lens:
  i. Separates parallel light rays.
  ii. Is thinner in the middle.
  iii. Is called a concave lens.

- The line that connects the centers of curvature for the faces of the lens is called the principal axis.

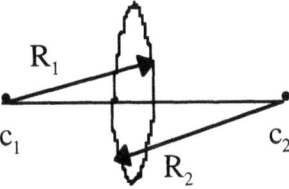

- There is a focal point associated with each face of a lens, and the distance from the lens to the focal point is the focal length of the lens.

- For a thin lens, the focal points are equidistant from the face of the lens.

- Just as for mirrors, the two math tools are that $\dfrac{1}{p} + \dfrac{1}{q} = \dfrac{1}{f}$ and

$$\text{linear magnification} = \text{l.m.} = \frac{\textit{linear dimension of image}}{\textit{corresponding linear dimension of object}} = \frac{q}{p}$$

- The standard rays for lenses (with the diverging case in parentheses) are the ray of light parallel to the principal axis is refracted so that it passes through (or appears to come from) the focal point of the lens.

- The ray of light that passes through (or would pass through) the focal point of the lens leaves the lens parallel to the principal axis.

- The ray of light that passes through the center of a lens is unchanged.

- Two examples are:

  A.

  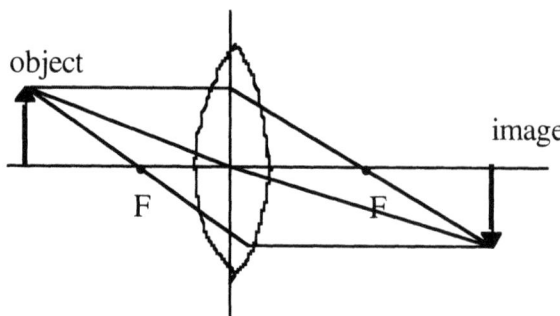

  The image is real, inverted, and magnified.

  B.

  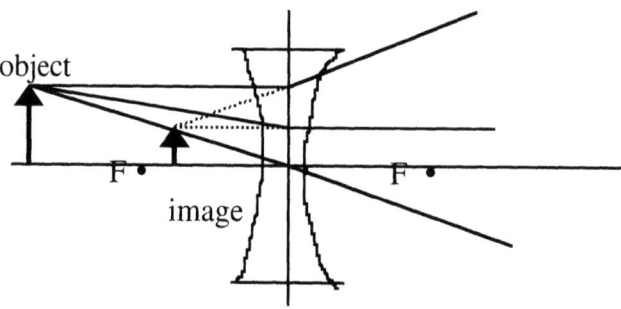

  The image is virtual, reduced, and erect.

- The sign convention for lenses is:
  1. The object distance is positive if the rays of light go from the object to the lens.

  2. The object distance is negative if the rays go from the lens to the object.

  3. The focal length of a converging lens is positive.

  4. The focal length of a diverging lens is negative.

  5. The image distance is positive if the rays of light go from the line to the image.

  6. If the rays of light go away from the image after leaving the lens the image distance is negative.

- The power of a lens, measured in diopters, is given by $P(\text{Diopters}) = \dfrac{1}{f(m)}$.

- Spherical lenses have two major defects:
  1. Spherical aberration where all rays parallel to the principal axis does <u>not</u> pass through the same point on the principal axis.

  2. Chromatic aberration, or color breakup, caused by the fact that different colors (wavelengths) of light have different indices of refraction.

- The fundamental idea in working a multiple lens problem is to just work each lens separately, i.e., the image of the first lens serves as the object for the second lens, and the image of the second lens serves as the object for the third lens, and so on.

- The total magnification for a lens system is the product of the magnification of each lens.

- Treating an eye as an optical instrument, myopia (nearsightedness) and hyperopia (farsightedness) are common defects of vision.

- Relative to eyes:
  1. The near point is the closest an object can be and be seen clearly.

  2. The far point is the furthest an eye can be and be seen clearly.

  3. The distance of most distinct vision is that distance for which all the detail of an object can be seen most clearly and distinctly, and is about 25 cm.

  4. The distance of most relaxed vision is infinity (or a long ways away).

  5. Accommodation is the ability of the eye to change its focal length.

- The myopic eye requires a diverging lens for it to see distant objects clearly.

- The hyperopic eye requires a converging lens to see up close objects clearly.

- A simple microscope is a converging lens where the object falls inside the focal point.

- Angular magnification is defined to be the angle subtended by the image divided by the angle subtended by the object, i.e., $m = \dfrac{\beta}{\alpha}$.

- For small angles $m = \dfrac{\beta}{\alpha} \cong \dfrac{\tan\beta}{\tan\alpha} \approx l.m.$

- In the case of a simple microscope, the angular magnification is given by
  $m = \dfrac{\tan\beta}{\tan\alpha} \cong \dfrac{q}{p}$

  a. If the image is at the distance of most distinct vision, (25 cm), $m = \dfrac{25cm}{p} = \dfrac{25cm}{f} + 1$

  [both p and f in centimeters]

  b. If the image is at the distance of most relaxed vision, the angular magnification is given by $m = \dfrac{25cm}{f}$, (f in centimeters)

- A compound microscope in its most basic form consists of two converging lenses as in the diagram.

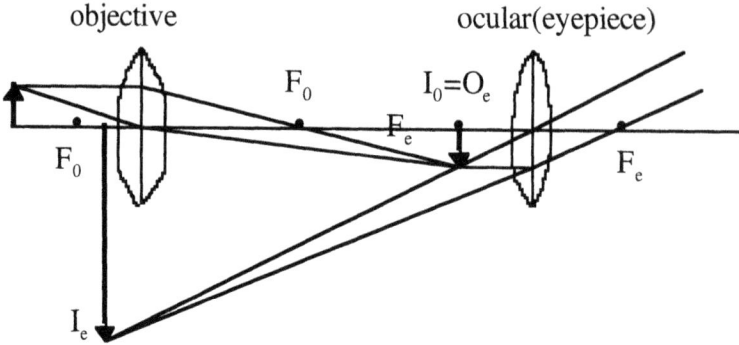

What the eye sees is an image that is virtual, inverted, magnified.

- An astronomical telescope in its most basic form consists of two converging lenses, the objective and ocular (eyepiece) that have the same basic ray diagram as the compound microscope. The image is virtual, inverted, magnified.

- If an astronomical telescope is used to look at a distant object, and if the final image is at infinity, the angular magnification is given by $m = \dfrac{f_o}{f_e}$ and the distance between the lenses is given by $f_o + f_e$.

- The Galilean telescope uses a diverging lens for the eyepiece and has an erect final image. The angular magnification is given by $m = \dfrac{f_o}{f_e}$.

- For two thin lenses in contact, the focal length of the combination is given by $\dfrac{1}{f_c} = \dfrac{1}{f_1} + \dfrac{1}{f_2}$.

- The lens-maker's equation says that $\dfrac{1}{f} = (n-1)\left(\dfrac{1}{R_1} - \dfrac{1}{R_2}\right)$ where f is focal length, n is the index of refraction of the lens material relative to the medium in which it is immersed, and R_1, R_2 are the radii of curvature of the surface of the lens. The sign convention says that R_1, R_2 are positive if the ray of light encounters a convex surface, and negative if the ray encounters a concave surface.

SOLVED PROBLEMS

▫▫ A converging lens of focal length 10 cm has a 5 cm tall object 15 cm in front of it. (a) Sketch the ray diagram and from the sketch describe the character of the image, and (b) find the height of the image.

(a)

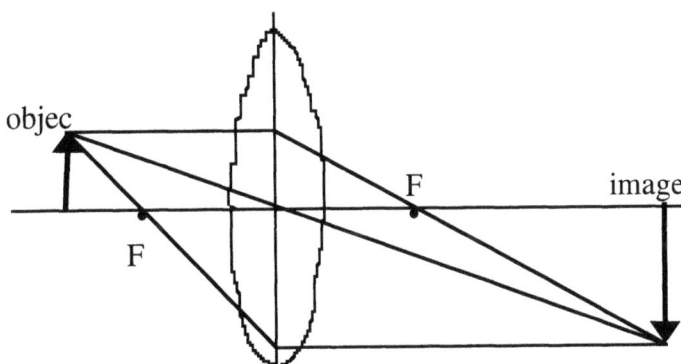

The image is real, inverted, and magnified.

(b) $\frac{1}{p} + \frac{1}{q} = \frac{1}{f} \Rightarrow \frac{1}{q} = \frac{1}{f} - \frac{1}{p} = \frac{1}{10cm} - \frac{1}{15cm}$

Thus, q = 30 cm. The linear magnification is thus $\frac{q}{p} = \frac{30cm}{10cm} = 3X$

Hence, l.m. = $\frac{height\ of\ image}{height\ of\ object} \Rightarrow 3 = \frac{ht\ of\ image}{5cm} \Rightarrow ht\ of\ image = 15cm$

▫▫ For the previous problem, suppose the object is 6 cm from the lens. (a) Sketch the ray diagram and from the sketch describe the character of the image, and (b) find the height of the image.

(a)

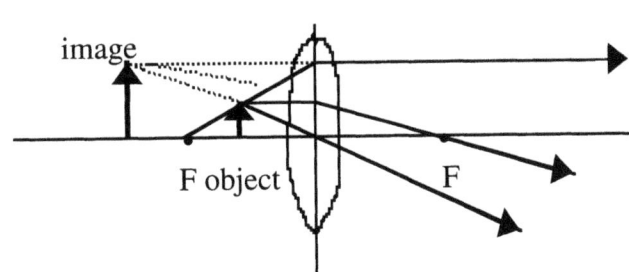

From the sketch, the image is erect, virtual and magnified.

(b) $\frac{1}{p} + \frac{1}{q} = \frac{1}{f} \Rightarrow \frac{1}{q} = \frac{1}{f} - \frac{1}{p} = \frac{1}{10cm} - \frac{1}{6cm}$. Thus, q = – 15 cm.

so that l.m. = $\frac{q}{p} = \frac{-15cm}{6cm} = 2.5X \Rightarrow l.m. = \frac{ht\ of\ image}{ht\ of\ object}$

⇒ ht of image = (l.m.)(ht of object)
= (2.5)(5 cm)
= 12.5 cm

☐☐ A diverging lens of focal length 15 cm is 12 cm from an object. (a) Sketch the ray diagram and from the sketch describe the character of the image, and (b) find the linear magnification of the lens.

(a)

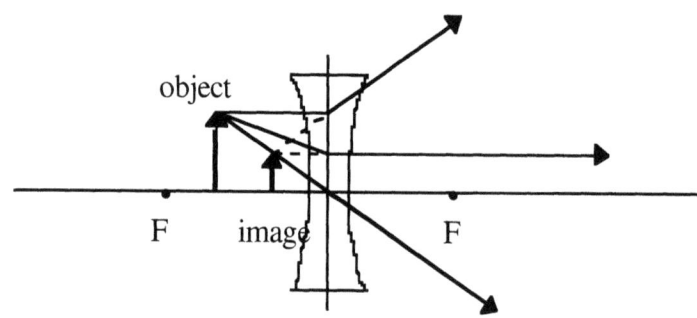

The image is erect, reduced, and virtual.

(b) $\frac{1}{p} + \frac{1}{q} = \frac{1}{f} \Rightarrow \frac{1}{q} = \frac{1}{f} - \frac{1}{p} = \frac{1}{-15cm} - \frac{1}{12cm}$

[Note: The focal length of a diverging lens is negative.]

Thus, q = - 6.67 cm.

l.m. $= \frac{q}{p} = \frac{-6.67cm}{12cm} = 0.556X$

☐☐ A simple microscope of focal length 12 cm is used to look at an object 5 cm from the lens. Find (a) where the image is formed and (b) the magnification of this microscope.

a) $\frac{1}{p} + \frac{1}{q} = \frac{1}{f} \Rightarrow \frac{1}{q} = \frac{1}{f} - \frac{1}{p} = \frac{1}{12cm} - \frac{1}{5cm}$, so that q = -8.57 cm.

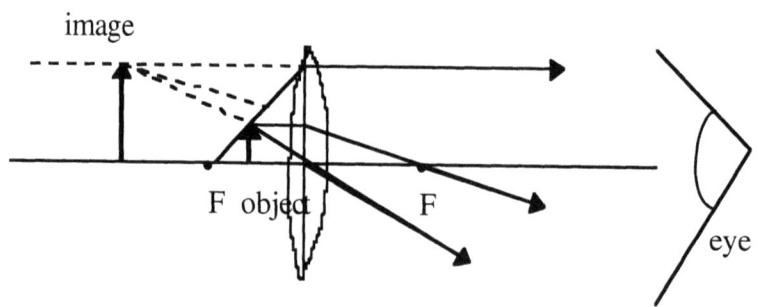

(b) l.m. = $\dfrac{q}{p} = \dfrac{-8.57cm}{5cm} = 1.71X$

☐☐ **For the previous problem, find the angular magnification if (a) the image is at the distance of most distinct vision, and (b) the distance of most relaxed vision.**

(a) For the distance of most distinct vision, 25 cm, m = $\dfrac{25cm}{f} + 1 = \dfrac{25cm}{12cm} + 1$

Thus m = 3.08X.

(b) For the distance of most relaxed vision, ∞, m = $\dfrac{25cm}{f} = \dfrac{25cm}{12cm} = 2.08X$

☐☐ **A 5 cm tall object is 12 cm from a converging lens of focal length 4 cm. 10 cm past the first lens is a converging lens of 2 cm. (a) Sketch the ray diagram for this system, (b) find the location of the final image, and (c) find the magnification of this system.**

(a)
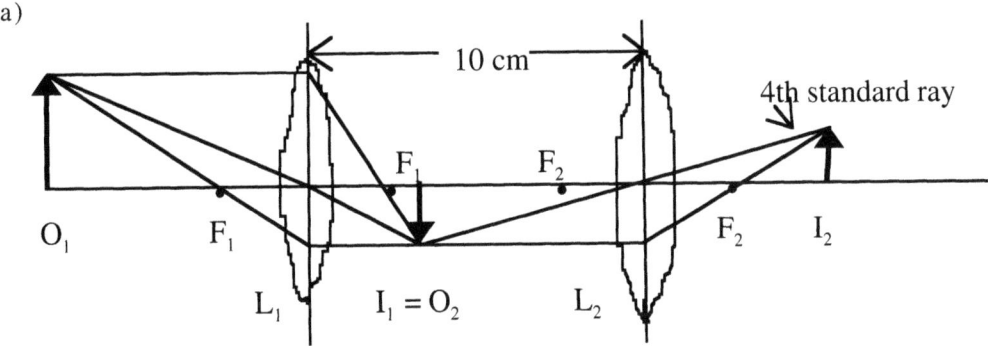

[Note. The 4rth standard ray is the following: Out of all of the rays coming from the first lens there has to be at least one that goes through the center of the 2nd lens and the tip of the first image.]

The final image from the sketch is real, erect, and reduced.

(b) For the first lens: $f_1 = 4$ cm, $p_1 = 12$ cm,

$\dfrac{1}{f_1} = \dfrac{1}{p_1} + \dfrac{1}{q_1} \Rightarrow \dfrac{1}{q_1} = \dfrac{1}{f_1} - \dfrac{1}{p_1} = \dfrac{1}{4cm} - \dfrac{1}{12cm}$

so that $q_1 = 6$ cm

For the second lens: $f_2 = 2$ cm, $p_2 = 4$ cm

$\dfrac{1}{q_2} = \dfrac{1}{f_2} - \dfrac{1}{p_2} = \dfrac{1}{2cm} - \dfrac{1}{4cm}$

so that $q_2 = 4$ cm

The final image is 4 cm from the second lens.

(c) $(l.m.)_T = (l.m.)_1 (l.m.)_2$

$= \dfrac{q_1}{p_1} \dfrac{q_2}{p_2} = \dfrac{6cm}{12cm} \dfrac{4cm}{4cm} = \dfrac{1}{2} X$

⬜⬜ An astronomical telescope has an objective lens of 150 cm and an ocular (eyepiece) lens of 6 cm. For a distant object, and when the final image is at the distance of most relaxed vision, find (a) the magnification, and (b) the distance between the objective and ocular lenses.

(a) For a distant object, when the final image is at ∞, m = $\dfrac{f_0}{f_e} = \dfrac{150cm}{6cm} = 25X$

(b) Here, the lenses are separated by a distance of $f_0 + f_e$ = 150 cm + 6 cm = 156 cm.

⬜⬜ The near point of an eye is 40 cm. What is the Power of a lens that would enable this eye to see print clearly at 20 cm?

The image of the corrective lens serves as the object for the eye. Thus, for the corrective lens, p = 20 cm, q = -40 cm, so that
$$\dfrac{1}{f} = \dfrac{1}{p} + \dfrac{1}{q} = \dfrac{1}{20cm} - \dfrac{1}{40cm} \Rightarrow f = 40cm$$

Since Power of a lens measured in kiopters is given by
$P(D) = \dfrac{1}{f(m)} \Rightarrow P = \dfrac{1}{0.4m} = 2.5D$

⬜⬜ The far point of an eye is 2 m. Find the Power of the lens that would enable this eye to see the stars clearly.

For the corrective lens, p = ∞ and q = - 2 m. Thus, $\dfrac{1}{f} = \dfrac{1}{p} + \dfrac{1}{q} \Rightarrow \dfrac{1}{f} = \dfrac{1}{\infty} - \dfrac{1}{2m}$
⇒ f = -2 m.
$P = \dfrac{1}{f(m)} = -0.5D$

⬜⬜ A compound microscope has an objective lens of 0.8 cm and an ocular (eyepiece) lens of focal length 4 cm. If the lenses are separated by a distance of 16 cm, find the magnification of this device if (a) the final image is at the distance of most distinct vision, and (b) if the final image is at the distance of most relaxed vision.

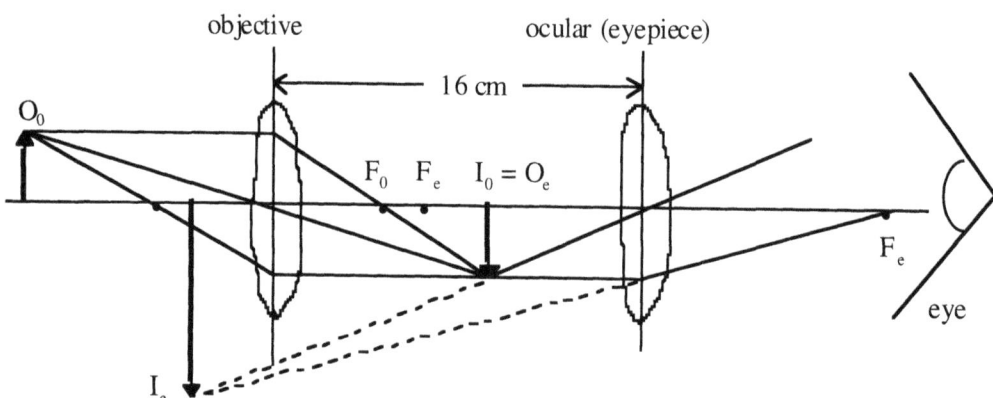

(a) For the compound microscope, the ocular lens acts as if it were a simple microscope. Thus, $m_T = m_0 m_e = \dfrac{q_0}{p_0} \dfrac{-25cm}{p_e}$.

For the eyepiece lens, $\dfrac{1}{f_e} = \dfrac{1}{p_e} + \dfrac{1}{q_e} \Rightarrow \dfrac{1}{p_e} = \dfrac{1}{f_e} - \dfrac{1}{q_e} = \dfrac{1}{4cm} + \dfrac{1}{25cm}$,

so that $p_e = 3.45$ cm. Hence, since the lenses are separated by 16 cm this means that $q_0 = 16$ cm - 3.45 cm = 12.55 cm.

For the objective lens, $\dfrac{1}{p_0} = \dfrac{1}{f_0} - \dfrac{1}{q_0} = \dfrac{1}{0.8cm} - \dfrac{1}{12.55cm} \Rightarrow p_0 = 0.854 cm$.

Thus, $m = \dfrac{12.55cm}{0.854cm} \dfrac{-25cm}{3.45cm} = 106X$.

(b) If the final image is at ∞, then for the eyepiece lens, $m_e = \dfrac{25cm}{f_e}$

$m_e = \dfrac{25cm}{4cm} = 6.25X$.

The image distance for the objective lens is 16 cm - 4 cm = 12 cm, since for the image of the ocular lens to be at infinity means the object for the eyepiece lens has to fall on the focal point of the eyepiece lens. Hence,
$\dfrac{1}{f_0} = \dfrac{1}{p_0} + \dfrac{1}{q_0} \Rightarrow \dfrac{1}{p_0} = \dfrac{1}{f_0} - \dfrac{1}{q_0} = \dfrac{1}{0.8cm} - \dfrac{1}{12cm}$

Thus, $p_0 = 0.857$ cm.

$m_0 = \dfrac{12cm}{0.857cm} = 14X$.

Finally, the total magnification if the final image is at infinity is given by
$m_T = m_0 m_e = (14)(6.25) = 87.5X$.

□□ A diverging lens of focal length -20 cm is to be used to form an image of the sun in order to study sunspots. Where should the screen be placed in order to focus the image of the sun?

One cannot place a virtual image on a screen, and a single diverging lens always forms an image that is virtual, reduced, and erect as can be seen from the sketch.

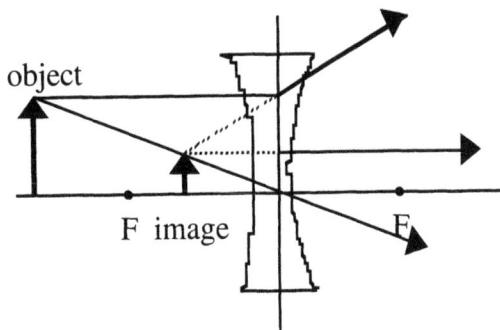

□□ A single converging lens has an object on the focal point. Where is the image formed?

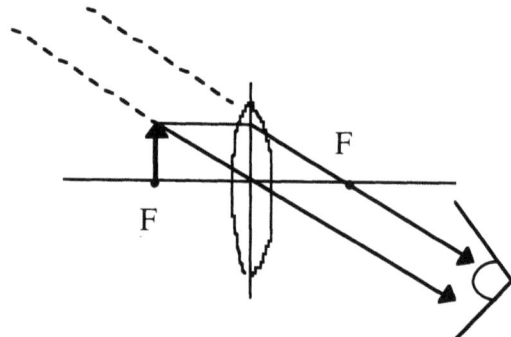

The image appears to be at infinity.

□□ Two thin lenses in contact have a combination focal length of 50 cm. One is a converging lens of focal length 20 cm, and the second lens is a diverging lens. Find the focal length of the diverging lens.

For two thin lenses in contact, $\frac{1}{f_c} = \frac{1}{f_1} + \frac{1}{f_2} \Rightarrow \frac{1}{50cm} = \frac{1}{20cm} + \frac{1}{f_2}$

Thus, $f_2 = -33.3$ cm or the focal length of the diverging lens is -33.3 cm.

□□ An object that is 4 cm tall is 25 cm from a converging lens of focal length 10 cm. 9 cm past the first lens is a diverging lens of focal length 8 cm. Find (a) the location of the final image of this lens system, and (b) the height of the image.

(a)

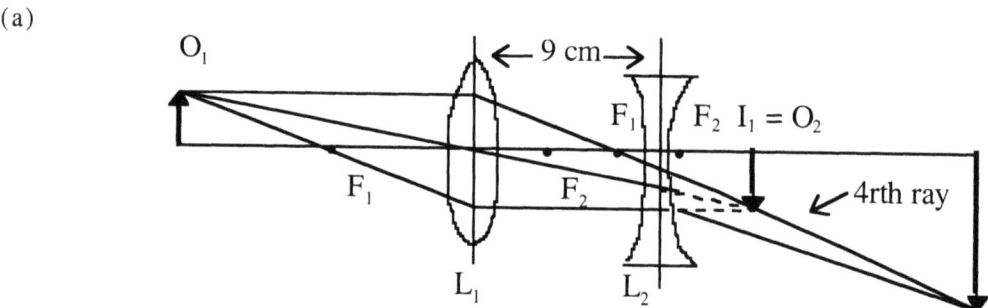

The rays will meet to form a real image that is inverted, real and magnified.

Lens # 1: $f_1 = 10$ cm
$\quad\quad\quad p_1 = 25$ cm

$\frac{1}{f_1} = \frac{1}{p_1} + \frac{1}{q_1} \Rightarrow \frac{1}{q_1} = \frac{1}{f_1} - \frac{1}{p_1} = \frac{1}{10cm} - \frac{1}{25cm} \Rightarrow q_1 = 16.7 cm$

Lens # 2: $f_2 = -8$ cm (it's a <u>diverging</u> lens)
$\quad\quad\quad p_2 = -7.7$ cm (rays go from lens to object)

Thus, $\dfrac{1}{q_2} = \dfrac{1}{f_2} - \dfrac{1}{p_2} = \dfrac{-1}{8cm} + \dfrac{1}{7.7cm}$

$q_2 = 205$ cm

Thus, the final image is 205 cm from the second lens.

(b) $(l.m.)_T = (l.m.)_1(l.m.)_2 = \dfrac{q_1}{p_1}\dfrac{q_2}{p_2} = \dfrac{16.7cm}{25cm}\dfrac{205cm}{-7.7cm}$

$(l.m.)_T = 17.8X$

Finally, $l.m. = \dfrac{ht\ of\ I}{ht\ of\ O} \Rightarrow ht\ of\ I = (l.m.)(ht\ of\ O)$

$= (17.8)(4\ cm) = 71.2$ cm

The height of the image of this system is 71.2 cm.

☐☐ A 35 mm slide has actual dimensions of 24 mm by 36 mm. It is desired to project an image of the slide onto a screen that is 7 m from the lens of the slide projector, and the dimensions of the screen are 1.2 m by 1.8 m. What is the focal length of the lens in the slide projector?

$l.m. = \dfrac{q}{p} = \dfrac{linear\ dimension\ of\ image}{corresponding\ linear\ dimension\ of\ object}$

Thus, $\dfrac{7m}{p} = \dfrac{1.2m}{2.4x10^{-3}m} \Rightarrow p = 1.4x10^{-2}m$

Hence, $\dfrac{1}{p} + \dfrac{1}{q} = \dfrac{1}{f} \Rightarrow \dfrac{1}{1.4x10^{-2}m} + \dfrac{1}{7m}$

$f = 1.397 \times 10^{-2}$ m

☐☐ At age 40, an individual needs a 2 diopter corrective lens to see print, that is 25 cm from the eye, clearly. At age 45, this individual, with the same lens, can see print clearly that is 30 cm from the eye. (a) Find the near point of the eye at age 40, and (b) find the near point of the eye at age 45.

(a) At age 40, q = the near point and the object distance = 25 cm.

Since $f(m) = \dfrac{1}{P(D)} \Rightarrow f(m) = \dfrac{1}{2}m = 50cm$

Thus, $\dfrac{1}{p} + \dfrac{1}{q} = \dfrac{1}{f} \Rightarrow \dfrac{1}{q} = \dfrac{1}{f} - \dfrac{1}{p} = \dfrac{1}{50cm} - \dfrac{1}{25cm}$, so that q = - 50 cm.

The near point of this eye at age 40 is 50 cm.

(b) At age 45, q = near point and p = 30 cm.

Hence, $\dfrac{1}{q} = \dfrac{1}{f} - \dfrac{1}{p} = \dfrac{1}{50cm} - \dfrac{1}{30cm} \Rightarrow q = -75cm$

At age 45, the near point is 75 cm. Note that the advance of the near point with age is a condition called presbyopia.

❏❏ A plano-convex lens has a flat surface and a curved surface that has a radius of curvature of 25 cm. Find (a) the focal length of this lens and (b) the Power of this lens. (The glass, which was used to make the lens has an index of refraction of 1.6) Hint: A flat surface has an infinite radius of curvature.

(a)

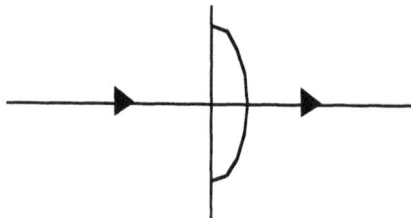

$$\frac{1}{f} = (n-1)\left(\frac{1}{R_1} - \frac{1}{R_2}\right) = (1.6-1)\left(\frac{1}{\infty} - \frac{1}{-25cm}\right)$$

The focal length of this lens is 41.7 cm. Note that it doesn't make any difference how the ray goes through the lens.

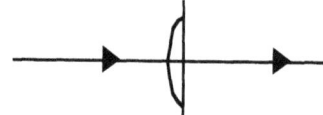

$$\frac{1}{f} = (1.6-1)\left(\frac{1}{25cm} - \frac{1}{\infty}\right) \Rightarrow 41.7cm \text{ as before.}$$

(b) Since Power (Diopters) = $\frac{1}{f(m)}$

Power = $\frac{1}{.417m}$ = 2.4D

❏❏ Consider the figure of the eye.

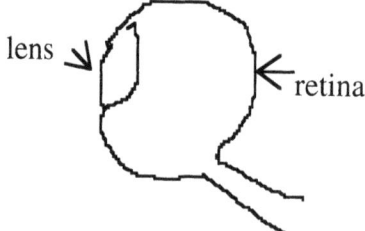

Of the three variables, focal length f, image distance q, and object distance p, (a) which is constant, and (b) what must the lens of the eye do to focus objects that are at different distances on the retina?

(a) The image distance q is constant.

(b) The lens of the eye must change its focal length for different object distances, as can be seen by looking at $\frac{1}{f} = \frac{1}{p} + \frac{1}{q}$. If q is constant, and p varies, then f must also vary.

☐☐ Consider the camera in the figure. Of the three variables, focal length f, image distance q, and object distance p, (a) which is constant, and (b) what must the camera do to focus objects that are at different distances from the film?

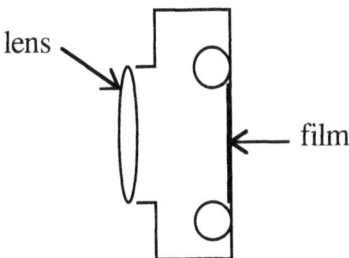

(a) For the camera, the focal length f is constant.

(b) The lens of the camera must be moved relative to the film, as can be seen by looking at $\frac{1}{p} + \frac{1}{q} = \frac{1}{f}$. If f is constant, and p varies, then q must vary. Thus for different object distances, the lens must be moved relative to the film to focus the image on the film.

☐☐ A Minox camera has a lens that has a focal length of 15 mm. What would be the dimensions of the image on the film of a page [8.5" x 11" or 21.6 cm x 27.9 cm] if it were photographed at a distance of 20 cm?

$$\frac{1}{f} = \frac{1}{p} + \frac{1}{q} \Rightarrow \frac{1}{q} = \frac{1}{f} - \frac{1}{p} = \frac{1}{1.5cm} - \frac{1}{20cm}$$

q = 1.62 cm.   l.m. = $\frac{q}{p} = \frac{1.62cm}{20cm} = 0.081X$

Since l.m. = $\frac{q}{p} = \frac{\text{width of image}}{\text{width of object}}$

⇒ width of image = (0.081)(21.6 m) = 1.75 cm
             = (0.081)(27.9 cm) = 2.26 cm

Hence, the dimensions of the image would be 1.75 cm x 2.26 cm.

☐☐ For the previous problem, suppose the next page was photographed from a distance of 30 cm. How far would the lens of the camera have to be moved to focus the image on the film?

Previously q = 1.62 cm for an object distance of 20 cm. Since,
$$\frac{1}{p} + \frac{1}{q} = \frac{1}{f} \Rightarrow \frac{1}{q} = \frac{1}{f} - \frac{1}{p} = \frac{1}{1.5cm} - \frac{1}{30cm}$$

q = 1.58 cm for p = 30 cm. Thus, the camera lens would have to be moved 1.62 cm - 1.58 cm = 0.04 cm = 0.4 mm, to focus the image on the film.

☐☐ **An object is placed 25 cm from a screen. Find the two points between the object and screen at which a 5 cm focal length lens may be placed to form an image on the screen.**

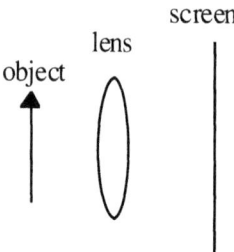

Let p = object distance, then q = image distance = 20 cm - p.

Thus, $\frac{1}{f} = \frac{1}{p} + \frac{1}{q} \Rightarrow \frac{1}{25cm} = \frac{1}{p} + \frac{1}{25cm - p}$, so that $\frac{1}{25cm} = \frac{(25cm - p) + p}{(25cm - p)p}$

then $(25cm)p - p^2 = 400 \text{ cm}^2$ <u>or</u>, $p^2 - (25cm)p + 400 \text{ cm}^2 = 0$

$$p = 6.9 \text{ cm or } 18.1 \text{ cm}$$

Thus, if the lens is 6.9 cm or 18.1 cm from the object an image will be formed on the screen.

☐☐ **A converging meniscus lens (see figure) has an index of refraction of 1.6, and the radii of curvature of its lenses 5 cm and 10 cm. (a) Find the focal length of this lens, and (b) if the lens is placed upward and filled with water, with the 10 cm radius next to the water, find the focal length of the water-glass lens combination. (The index of refraction of water is 1.33)**

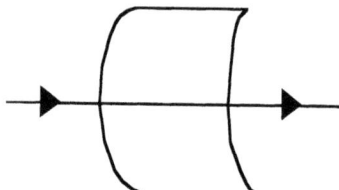

converging meniscus lens has to be
thicker in the middle

(a) From the lens-maker's equation,

$\frac{1}{f} = (n-1)\left(\frac{1}{R_1} - \frac{1}{R_2}\right)$ (both R_1 and R_2 are positive because the surfaces encountered are both convex.)

$\frac{1}{f} = (1.6 - 1)\left(\frac{1}{5cm} - \frac{1}{10cm}\right) \Rightarrow \frac{1}{f_{glass}} = (0.6)\left(\frac{1}{10}\right)$, so that $f_g = 16.7$ cm.

(b) For the water-glass lens:
R₁ = 10 cm
R₂ = ∞

$$\frac{1}{f_{water}} = (1.33 - 1)\left(\frac{1}{10cm} - \frac{1}{\infty}\right)$$

[Note that the radius of curvature of a flat surface is infinity.]

$$\frac{1}{f_{water}} = \left(\frac{1}{3}\right)\left(\frac{1}{10cm}\right) \Rightarrow f_{water} = 30cm$$

For the combination, $\frac{1}{f_c} = \frac{1}{f_1} + \frac{1}{f_2} = \frac{1}{16.7cm} + \frac{1}{30cm}$

The focal length of the combination is 10.7 cm.

▫▫ **An object is placed in front of a converging lens, and the image is formed 20 cm from the lens on a screen. A diverging lens is now placed halfway between the converging lens and the screen and it is found that the screen must be moved 25 cm further away from the converging lens to focus the image on the screen. Find the focal length of the diverging lens.**

First: (converging lens)

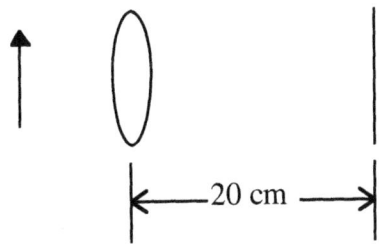

Second : (converging lens + diverging lens)

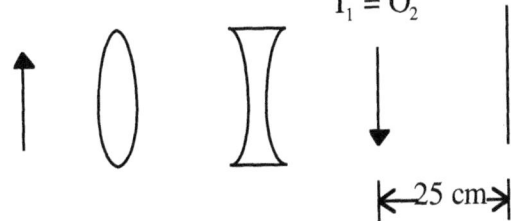

For the diverging lens, q = 35 cm and p = - 10 cm (because the rays go from the lens to the object).
Thus, $\frac{1}{f} = \frac{1}{p} + \frac{1}{q} = \frac{-1}{10cm} + \frac{1}{35cm} \Rightarrow f = -14cm$.

The focal length of the diverging lens is –14 cm.

❐❐ **Where is the near point of an eye for which a 2 D lens is prescribed? (Assume the corrective lens allows this eye to see print clearly at 25 cm.)**

$P(D) = \frac{1}{f(m)} \Rightarrow f = \frac{1}{2}m = 50cm$

Since the image of the corrective lens serves as the object for the eye, p = 25 cm.

$\frac{1}{p} + \frac{1}{q} = \frac{1}{f} \Rightarrow \frac{1}{25cm} + \frac{1}{q} = \frac{1}{50cm}$

$\frac{1}{q} = \frac{1}{50cm} - \frac{1}{25cm} \Rightarrow q = -50cm$

The near point of the eye is 50 cm.

❐❐ **Find the far point of an eye for which the corrective lens has a Power of -0.5 D.**

The image of the corrective lens serves as the object for the eye. If this eye is to see clearly an object that is a long ways away, $p \to \infty$, $f = \frac{1}{P(D)} = -2m$.

$\frac{1}{p} + \frac{1}{q} = \frac{1}{f} \Rightarrow \frac{1}{\infty} + \frac{1}{q} = -\frac{1}{2}m$ so that q = - 2m.

The far point of the eye is 2 m.

WAVE OPTICS I

THE FUNDAMENTAL IDEAS ARE THE FOLLOWING:

- Two light waves that are as indicated in the figure are said to be "in phase".

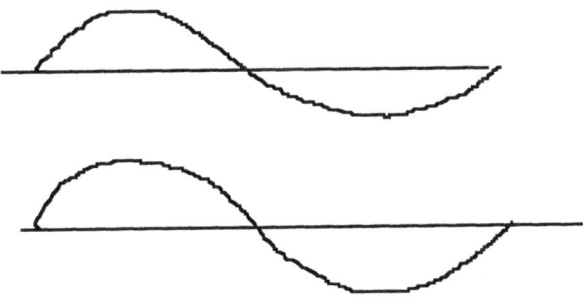

- Two light waves as in the figure are said to be "out of phase" by an amount $\frac{\lambda}{2}$, (or π radians or 180°).

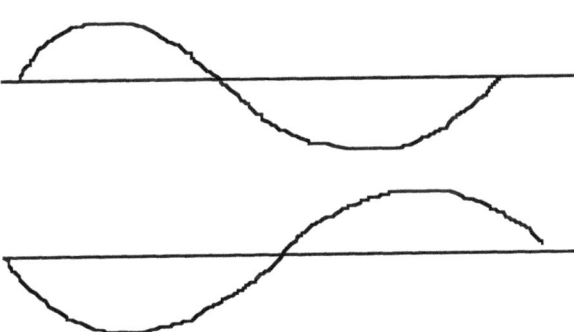

- Constructive interference is the addition of two waves that are in phase into a single wave, i.e., a region of light.

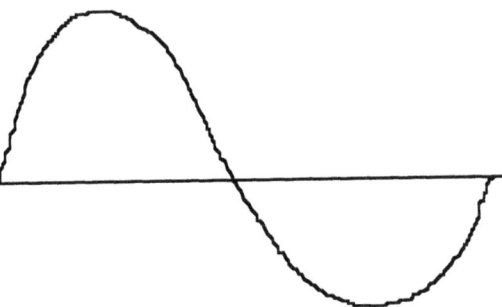

291

- Destructive interference is the addition of two waves that are out of phase by an amount $\frac{\lambda}{2}$, to get a region of dark.

- Path difference is defined to be how much further one wave traveled than did another.

- When light is reflected from a material that has a higher index of refraction there is a phase shift of $\frac{\lambda}{2}$.

- When light is reflected from a surface that has no phase shift there is no phase shift.

- The difference in phase shift between two waves is called phase shift difference.

- Total difference is defined to be path difference plus phase shift difference.

- The condition for constructive interference is that two (or more) waves have a total difference of $0, 1\lambda, 2\lambda, \ldots m\lambda$; $m = 0,1,2,3,\ldots$.

- The condition for destructive interference is that two (or more) waves have a total difference of $\frac{\lambda}{2}, \frac{3\lambda}{2}, \frac{5\lambda}{2}, \ldots \left(m+\frac{1}{2}\right)\lambda$; $m = 0,1,2,3,\ldots$.

- In 1801, Thomas Young demonstrated the wave-like nature of light with a double slit experiment as in the figure:

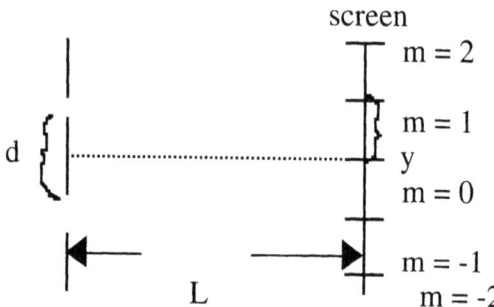

    The condition for bright bands is that the path difference between two waves is $m\lambda = d\sin\theta = \frac{yd}{L}$, $m = 0,1,2,\ldots$. m is said to be the order number. The $m=0$ bright band is referred to as the central bright fringe. The condition for dark bands is that the path difference between two waves is $\left(m+\frac{1}{2}\right)\lambda = d\sin\theta = \frac{yd}{L}$. Note that the double slit pattern is a path difference phenomenon.

- Huygen's Principle says that every point of a wave front may be considered the source of secondary wavelets that spread out in all directions with a speed equal to the speed of propagation of the wave.

- Thin films use a coating on a lens (or piece of glass) in order to have the reflected waves undergo either constructive or destructive interference.

- The # 1 and # 2 waves will undergo either constructive or destructive interference depending upon the thickness of the film and/or the indices of refraction of the materials involved.

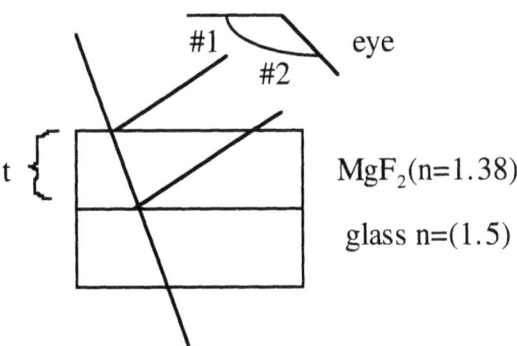

- From Snell's Law, $n_{material} = \dfrac{\lambda_{air\ or\ vacuum}}{\lambda_{material}}$, thus, $\lambda_{material} = \dfrac{\lambda_{air}}{n_{material}}$

SOLVED PROBLEMS

☐☐ **Two radio transmitters are used to calibrate a third as in the figure. What is the path difference between the two signals?**

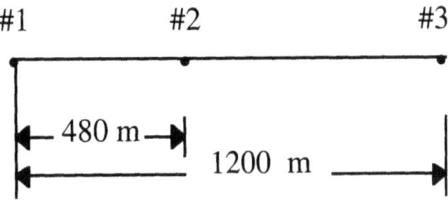

The path difference between the signals from the #1 and #2 transmitters is 1200 m - 480 m = 720 m.

☐☐ **Suppose that in the previous problem the #1 and #2 transmitters send out signals that are in phase. If the wavelength of the signal is 6 m, will there be constructive or destructive interference at the #3 transmitter?**

Since λ = 6 m, and path difference is 720 m,

path difference = 720 m $\left(\dfrac{1\lambda}{6m}\right) = 120\lambda$

Thus, there is constructive interference between the signals from the #1 and #2 transmitters.

☐☐ **In a Young's double slit experiment, the distance from the two slits to the screen is 2.97 m, and the distance from the central bright maximum to the third bright fringe is 1.6 cm. If the distance between the two slits is 0.35 mm, find the wavelength of the light.**

For bright bands, $m\lambda = \dfrac{yd}{L}$.

Hence, $3\lambda = \dfrac{(1.6 \times 10^{-2} m)(0.35 \times 10^{-3})}{2.97 m} \Rightarrow \lambda = 6.29 \times 10^{-7} m = 629 nm$

☐☐ Consider the figure: A thin soap film is in air. Find the phase shift difference between the #1 and #2 reflected waves.

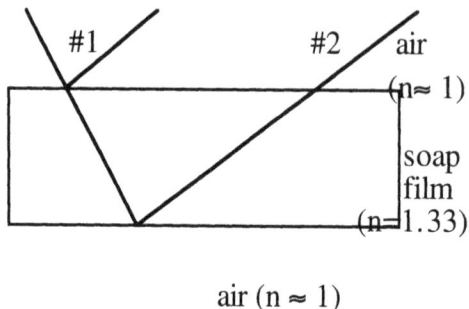

The #1 wave is reflected from a material (soap) that has a higher index of refraction, thus phase shift #1 = $\frac{\lambda}{2}$. The #2 wave is reflected from a material (air) that has a lower index of refraction, thus phase shift #2 = 0. Therefore, the phase shift difference is $\frac{\lambda}{2} - 0 = \frac{\lambda}{2}$.

☐☐ For the previous problem, assume that the incident ray is almost perpendicular to the surface and that the thickness of the soap film is t. Find the path difference between the two reflected waves.

The #2 wave travels down a thickness t and back up a thickness t that the #1 wave doesn't have to travel. Thus the path difference between the two waves is 2t.

☐☐ A double slit device produces an interference pattern on a screen that is 3.7 m away. If the third dark band is 1.3 cm away from the central bright fringe, find the wavelength of the light if the distance between the slits is 0.4 mm.

For dark bands, $\left(m + \frac{1}{2}\right)\lambda = d \sin\theta = \frac{yd}{L}$.

Thus, for the third dark band,

$2.5\lambda = \frac{(1.3 \times 10^{-2} m)(4 \times 10^{-4} m)}{3.7 m}$

$\lambda = 5.62 \times 10^{-7}$ m

$\lambda = 5.62$ nm

☐☐ Newton's rings are observed when a curved piece of glass is placed over a flat piece of glass as in the figure. When viewed from above the eye sees a series of light and dark bands. Show that where the pieces of glass touch there is a dark band.

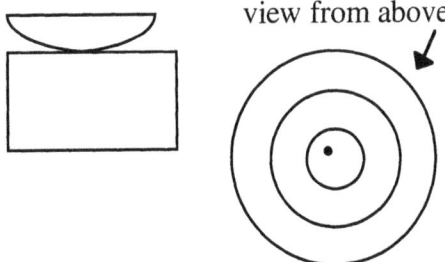

view from above

This is essentially a film of air. Where the pieces of glass touch the air film is almost of zero thickness. Thus, the path difference between the #1 and #2 waves is very close to zero. The #1 wave is reflected from the upper surface of the film of air, while the #2 wave is reflected from the lower surface of the air film. Thus, the phase shift difference between the two waves is p.s. #2 = $\frac{\lambda}{2}$, p.s. #1 = 0 ⇒ phase shift difference = $\frac{\lambda}{2}$. The total difference is path difference + phase shift difference, so that total difference = $\approx 0 + \frac{\lambda}{2} \cong \frac{\lambda}{2}$. Destructive interference (dark band) thus occurs where the pieces of glass touch.

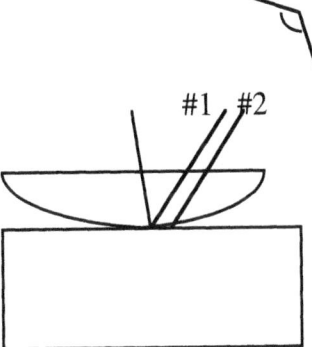

☐☐ **For the previous problem, find an expression that gives the thickness of the air wedge at the nth bright band.**

Let the thickness of the air wedge at the nth bright band be t. From the previous problem the phase shift difference between the #1 and #2 waves is $\frac{\lambda}{2}$.

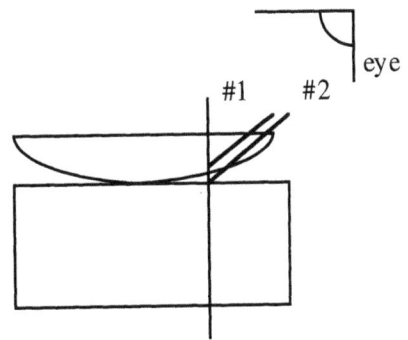

Since the condition for constructive interference is that the total difference is $m\lambda$, then $m\lambda = 2t + \dfrac{\lambda}{2} \Rightarrow \dfrac{\lambda\left(m - \dfrac{1}{2}\right)}{2} = t$.

If $m = 0 \Rightarrow t = \dfrac{-\lambda}{4}$ which is impossible.

If $m = 1 \Rightarrow t = \dfrac{\lambda}{4}$,

If $m = 2 \Rightarrow t = \dfrac{3\lambda}{4}$,

If $m = 3 \Rightarrow t = \dfrac{5\lambda}{4}$.

Thus, the thickness of the air wedge at the nth bright band is $t = \dfrac{(2n-1)\lambda}{4}$, where $n = 1, 2, 3, \ldots$.

▫▫ Two pieces of glass have a piece of paper between them at one end and touch at the other end as in the figure. Find the expression that gives the thickness of the air wedge at the nth dark band, measured from where the pieces of glass touch.

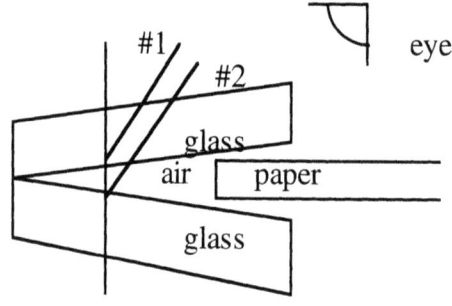

Let t be the thickness of the air wedge. Then the path difference between the #1 and #2 waves is 2t. The phase shift for the #1 wave is 0, because it is reflected from the upper surface of the air wedge. The phase shift for the #2 wave is $\dfrac{\lambda}{2}$, because it is reflected from the surface of the lower piece of glass. Thus the phase shift difference between the two waves is $\dfrac{\lambda}{2}$. The condition for destructive interference is $\left(m + \dfrac{1}{2}\right)\lambda$ which has to equal a total difference of $2t + \dfrac{\lambda}{2}$. Hence, for dark bands to occur in the reflected light, $\left(m + \dfrac{1}{2}\right)\lambda = 2t + \dfrac{\lambda}{2} \Rightarrow m\lambda = 2t \Rightarrow t = \dfrac{m\lambda}{2}$.

If $m = 0 \Rightarrow t = 0$ (where pieces of glass touch)

If $m = 1 \Rightarrow t = \dfrac{\lambda}{2}$

If $m = 2 \Rightarrow t = \lambda$

If $m = 3 \Rightarrow t = \dfrac{3\lambda}{2}$

Thus, the thickness of the air wedge at the nth dark band measured from where the pieces of glass touch is $t = \dfrac{n\lambda}{2}$, for n = 1,2,3,....

☐☐ **A coated lens appears to reflect yellow light (λ = 590 nm). Find the minimum thickness of the magnesium fluoride coating that would enable the yellow light to undergo constructive interference. (Assume that the eye is at almost normal incidence.)**

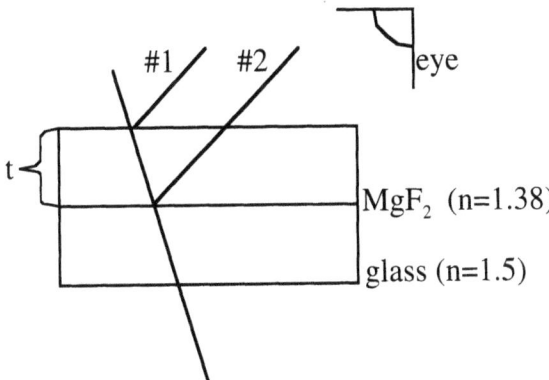

Here the path difference between the #1 and #2 waves is 2t. The phase shift for both the #1 and #2 waves is $\dfrac{\lambda}{2}$. Hence, the phase shift difference is 0. For constructive interference, the total difference equals $m\lambda$. Thus $m\lambda = 2t + 0 \Rightarrow t = \dfrac{m\lambda}{2}$.

The minimum thickness occurs when m = 1.

$t_{min} = \dfrac{\lambda_{mat}}{2} = \dfrac{\lambda_{air}}{2n} = \dfrac{590nm}{2(1.38)} \cong 214nm$

Note that the wavelength of light in the magnesium fluoride coating is not the same as the wavelength in air.

☐☐ **Two narrow slits are separated by 0.35 mm. Light of wavelength 480 and 650 nm is used to form a pattern on screen that is 2 m from the slits. Find the separation distance between the 2nd order bright fringes on the screen.**

For constructive interference $m\lambda = d\sin\theta = \dfrac{yd}{L}$.

(a) For the 480 nm light: $2(480 \times 10^{-9} m) = \dfrac{y_2(.35 \times 10^{-3} m)}{2m}$

y_2 = 5.49mm = 5.49 x 10^{-3} m

(b) For the 650 nm light: $2(650 \times 10^{-9} m) = \dfrac{y_2(.35 \times 10^{-3} m)}{2m}$

y_2 = 7.43 mm = 2.43 x 10^{-3} m

Thus, the 2nd order bright lines are separated by 7.43 x 10^{-3} m - 5.49 x 10^{-3} m = 1.94 x 10^{-3} m or the 2nd order bright lines are separated by 1.94 mm.

❏❏ **Two glass plates of length 10 cm are separated by a very fine wire at one end, and touch at the other, forming an air wedge. If there are 10 dark fringes per centimeter, estimate the diameter of the wire. Assume that light of wavelength 590 nm is incident normally from above.**

The thickness of the air wedge at the nth dark band (counting from where the pieces of glass touch) is given by $t = \dfrac{n\lambda}{2}$. Since there are approximately 100 dark fringes, at the place where the wire is between the pieces of glass, the thickness of the air wedge is given by $t = \dfrac{(100)590nm}{2} = 2.95 \times 10^{-5} m$. Thus, the approximate diameter of the wire is about 2.95×10^{-5} m.

❏❏ **Newton's rings are formed by placing a plano-convex lens on an optically flat piece of glass. Find an expression that gives the radius of curvature of the convex surface of the lens as a function of the wavelength of the light, the thickness of the air wedge at the nth bright circle, and the radius of the nth bright circle.**

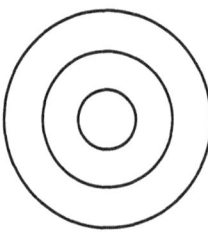

Newton's rings when viewed from above.

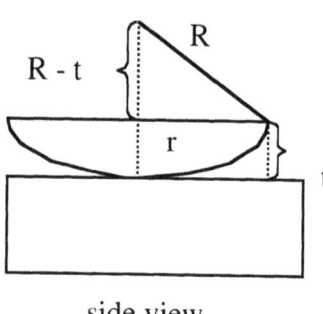

side view

R = radius of curvature of lens.

r = radius of the nth bright circle.

T = thickness of air wedge at n^{th} bright band

From the sketch, using the Pythagorean Theorem: $(R - t)^2 + r^2 = R^2$

$R^2 + t^2 - 2Rt + r^2 = R^2$

Thus, since t is very small, $t^2 \approx 0$, so that $r^2 = 2Rt$

$R \cong \dfrac{r^2}{2t}$

☐☐ **In a Newton's rings experiment, the plano-convex lens has a radius of curvature of 2.5 m. If light of wavelength 590 nm is used, what is the number of bright circles observed within a radius of 1 cm?**

From the previous problem, the thickness of the air wedge at the bright circle that has a radius of 1 cm is given by $\frac{r^2}{2R} = t = \frac{(1 \times 10^{-2} m)^2}{2(2.5m)} = 2 \times 10^{-5} m$.

Since the thickness of the air wedge at the nth bright circle is given by
$t = \frac{(2n-1)\lambda}{4} \Rightarrow \frac{4t}{\lambda} = 2n - 1$.

Thus, $\frac{4(2 \times 10^{-5} m)}{590 \times 10^{-9} m} = 2n - 1 \cong 135$, so that $2n = 136 \Rightarrow n = 68$. Finally, there are 68 bright circles observed within a radius of 1 cm.

☐☐ **A MgF_2 coating (n = 1.38) on a glass lens (n = 1.5) is 8.3×10^{-4} mm thick. White light is incident normally on the lens. Which wavelengths of visible light (400-700 nm) are missing in the reflected light?**

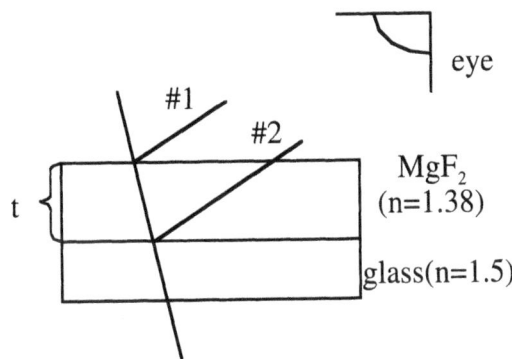

For light to be missing, it must have undergone destructive interference. Thus, the total difference (path difference plus phase shift difference) must equal $(m + 1/2)\lambda$. The path difference between the two waves is 2t. The phase shift difference is 0. Hence,
$(m + 1/2)\lambda = 2t$
$\lambda_{mat} = \frac{2t}{m + \frac{1}{2}} \Rightarrow \lambda_{air} = \frac{2tn}{m + \frac{1}{2}}, m = 0, 1, 2...$

since $\lambda_{mat} = \frac{\lambda_{air}}{n}$.

Finally, the wavelengths for which destructive interference will take place are given by
$\lambda_{air\ m=0} = \frac{2(8.3 \times 10^{-7} m)(1.38)}{\frac{1}{2}} = 4.58 \times 10^{-6} m$, <u>not</u> visible light.

$\lambda_{air\ m=1} = \frac{2(8.3 \times 10^{-7} m)(1.38)}{1.5} = 1.52 \times 10^{-6} m$, <u>not</u> visible light.

$$\lambda_{air\ m=2} = \frac{2(8.3x10^{-7}m)(1.38)}{2.5} = 9.16x10^{-7}m, \text{ not visible light.}$$

$$\lambda_{air\ m=3} = \frac{2(8.3x10^{-7}m)(1.38)}{3.5} = 6.55x10^{-7}m = 655nm.$$

$$\lambda_{air\ m=4} = \frac{2(8.3x10^{-7}m)(1.38)}{4.5} = 5.09x10^{-7}m = 509\ nm.$$

$$\lambda_{air\ m=5} = \frac{2(8.3x10^{-7}m)(1.38)}{5.5} = 4.17x10^{-7}m = 417nm.$$

$$\lambda_{air\ m=6} = \frac{2(8.3x10^{-7}m)(1.38)}{6.5} = 3.52x10^{-7}m, \text{ not visible light.}$$

Thus, the wavelengths 655, 509, and 417 nm will be missing from the reflected light.

☐☐ **Two slits are separated by 0.4 mm and have monochromatic light of wavelength 590 nm incident upon them. The pattern is viewed on a screen 1.5 m from the slits. How many bright fringes are there between the central bright fringe and an angle of 2 °?**

For bright bands, $m\lambda = d\sin\theta$

Thus, $m = \frac{(4x10^{-4}m)(\sin 2°)}{590x10^{-9}m} = 23.7$, or there are 23 bright fringes between the central bright fringe and 2°.

☐☐ **A MgF$_2$ coating (n = 1.38) is on glass (n = 1.6). Find the three minimum thicknesses that will enable the eye to see yellow light (λ = 590 nm) in the reflected light. Assume the light is incident normally on the glass.**

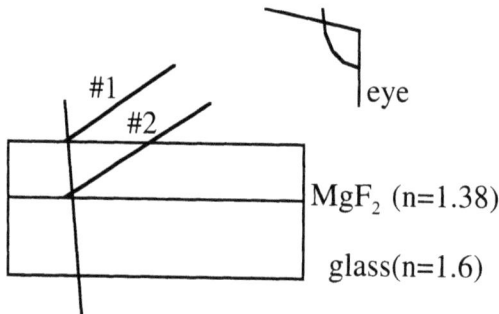

In order for the eye to see yellow light, constructive interference must take place. Thus, the total difference must equal $m\lambda$. The path difference between the 2 waves is 2t, and the phase shift difference is 0. Therefore,

$m\lambda = 2t \Rightarrow t = \frac{m\lambda_{mat}}{2} = \frac{m\lambda_{air}}{2n}$.

If $m = 0 \Rightarrow t = 0$, physically impossible.

If $m = 1 \Rightarrow t = \dfrac{(1)(590nm)}{2(1.38)} \approx 214nm$

If $m = 2 \Rightarrow t = 428$ nm

If $m = 3 \Rightarrow t = 642$ nm

So that the three minimum thicknesses of the coating that will provide constructive interference for yellow light are 214, 428, and 642 nm.

⊡⊡ **A double slit experiment is performed using light of wavelength 550 nm. If the distance between the slits is 5.3 x 10^{-5} m, find the angular separation between the second and third dark bands, measured from the central bright fringe.**

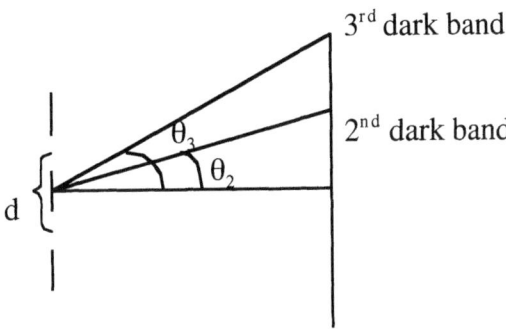

For destructive interference $\left(m + \dfrac{1}{2}\right)\lambda = d \sin\theta, \; m = 0,1,2,...$

For the third dark band, $\dfrac{2.5\lambda}{d} = \sin\theta_3$

$\sin\theta_3 = \dfrac{(2.5)(550 x 10^{-9} m)}{5.3 x 10^{-5} m}$

$\sin\theta_3 = 2.59 x 10^{-2} \Rightarrow \theta_3 = 1.49°$

For the second dark band, $\dfrac{1.5\lambda}{d} = \sin\theta_2 = \dfrac{1.5(550 x 10^{-9} m)}{5.3 x 10^{-5} m} \Rightarrow \sin\theta_2 = 1.56 x 10^{-2}$

$\theta_2 = 0.89°$

Thus, the angular separation between the third and second dark bands is $\theta_3 - \theta_2$
$1.49° - 0.89° = 0.6°$.

☐☐ Light of wavelength 661 nm is incident normally on a thin layer of gasoline (n = 1.4) that is floating on the surface of water (n = 1.33). If the gasoline layer is 2.36 x 10⁻⁷ m thick, will the reflected light undergo constructive or destructive interference?

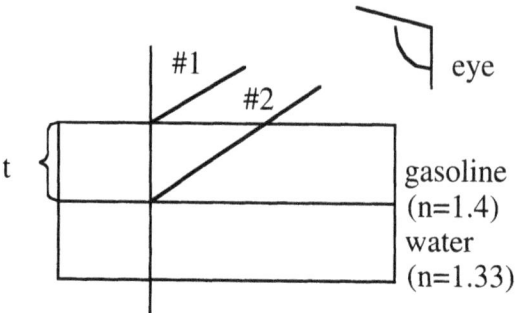

Here the total difference between the two waves is given by the path difference plus the phase shift difference. If the total difference is 0, 1λ, 2λ, 3λ, ..., the #1 and #2 waves will undergo constructive interference, and if the total difference is $0, \frac{\lambda}{2}, \frac{3\lambda}{2}, \frac{5\lambda}{2}, ...$ the #1 and #2 waves will undergo destructive interference. The phase shift difference is $\frac{\lambda}{2}$, thus the total difference = $2(2.36 \times 10^{-7} m) + \frac{\lambda}{2}$. For the light in the layer of gasoline,

$$\lambda_{gasoline} = \frac{\lambda_{air}}{n} = \frac{661 nm}{1.4} = 472 nm \Rightarrow t = \frac{\lambda_{gasoline}}{2}$$

Thus, in terms of wavelengths, the total difference is $2\left(\frac{\lambda}{2}\right) + \frac{\lambda}{2} = \frac{3\lambda}{2}$, a condition for destructive interference.

☐☐ A soap film is stretched over a wire framework as in the figure, and when the light reflected from the film is viewed a series of light and dark bands are formed. If the film is held so that it is in the vertical position, the light and dark bands move down and the bands become bands that have the colors of the rainbow. Explain why.

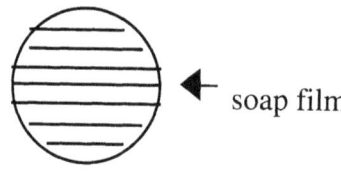

As the film is held in the vertical position, the pull of gravity causes the film to become thinner at the top, thus moving the position of the light and dark bands. The colors are caused because different wavelengths have different thicknesses for constructive interference. This is the reason that after a rain, puddles of water that have an oil film on them in a parking lot exhibit the colors of the rainbow.

WAVE OPTICS II

THE FUNDAMENTAL IDEAS ARE THE FOLLOWING:

- Diffraction is the bending of light waves as they pass an obstacle.

- For a single-slit diffraction pattern the dark fringes are given by $m\lambda = a\sin\theta = \frac{ay}{L}$, where a is the width of the slit, y is the distance from the central bright fringe, and L is the distance from the slit to the screen.

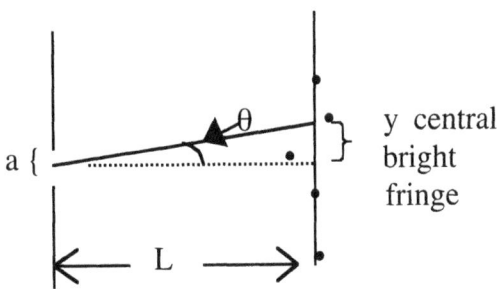

- The bright fringes are given approximately by $\left(m + \frac{1}{2}\right)\lambda = a\sin\theta = \frac{ay}{L}$.

- The ability of any optical system to produce sharp images is called its resolution. The resolution of any optical system is limited by diffraction.

- For a circular aperture, the first minimum in the diffraction pattern is given by $a\sin\theta = 1/22\lambda$, where a is the diameter of the circular aperture.

- Rayleigh's criterion says that two objects are just resolved (able to be seen as separate objects) when the central maximum of one diffraction pattern coincides with the first minimum of the second pattern.

- For small angles measured in radians, $\sin\theta \approx \theta$. Thus, $\theta = \frac{1.22\lambda}{a}$, where λ is wavelength and a is the diameter of the circular aperture. But from the definition of $\theta(rad) = \frac{s}{r}$, i.e. (arc length)/(radius), the angular separation so that objects can be seen as separate objects is thus given by $\theta = \frac{1.22\lambda}{a} = \frac{s}{r}$.

- A grating consists of many very fine slits cut into a glass plate. When light is passed through the grating (in this case a transmission grating, constructive and destructive interference takes place.

- The expression that gives the positions of the places where constructive interference takes place is $m\lambda = d\sin\theta$, where m is called order number, d is the distance between the slits, and θ is the angle between the bright band and the central bright fringe.

- If the incident light is monochromatic, a series of light and dark bands will be formed on a screen.

- If the light is polychromatic, then bright bands that repeat can be formed, in which case the lines are said to comprise a bright spectra. If the constructive interference is continuous, i.e., The colors change from red to yellow to blue without any dark bands in between the spectrum is said to be continuous.

SOLVED PROBLEMS

❐❐ **Light of wavelength 590 nm is incident normally on a slit of width .05mm. The pattern is observed on a screen 1.8 m away. Find (a) the width of the central bright fringe and (b) the distance between the first and second outer minima.**

(a) For the single slit diffraction pattern, the distance between the m = +1 and m = -1 minima is the width of the central bright fringe. Thus, for minima, $m\lambda = a\sin\theta = \frac{ay}{L}$.

Thus, $\frac{(1)\lambda L}{a} = y_1 = \frac{(1)(590 \times 10^{-9} m)(1.8m)}{0.05 \times 10^{-3} m} = 0.0212m$ or $y_1 = 2.12 cm \Rightarrow y\text{-}1 = 2.12 cm.$

So the width of the central bright fringe is 4.24 cm.

(b) The distance between the second and first order minima is just $y_2 - y_1$. Since $y_1 = 2.12 cm$, $y_2 = \frac{(2)\lambda L}{a} = \frac{2(590 \times 10^{-9} m)(1.8m)}{0.05 \times 10^{-3} m} = 0.0425m = 4.25cm$. Thus, the distance between the second and first order minima on the screen is $y_2 - y_1 = 4.25$ cm - 2.12 cm = 2.13 cm.

❐❐ **Find the minimum separation between two objects on the moon (3.84 x 10⁸ m from earth) for a telescope of diameter 4.5 m to see them as separate objects. (Assume λ = 550 nm.)**

From the Rayleigh criterion, $\theta = \frac{1.22\lambda}{a} = \frac{s}{r} \Rightarrow \frac{1.22(550 \times 10^{-9} m)}{4.5m} = \frac{s}{3.84 \times 10^8 m}$

s = 57.3 m

Thus, for light of wavelength 550 nm the minimum separation for two objects on the moon to be resolved with this telescope is 57.3 m.

❐❐ **A diffraction grating has 650 lines/mm. For light of wavelength 632 nm, what is the angular separation between the first and second order maxima?**

Here $m\lambda = d\sin\theta$, and $d = \frac{1}{650} \times 10^{-3} m$

For the first order: $\frac{(1)(632 \times 10^{-9} m)}{\frac{1}{650} \times 10^{-3} m} = \sin\theta_1 \cong 0.41$

$\Rightarrow 24.3° = \theta_1$

For the second order: $\dfrac{(2)(632 \times 10^{-9} m)}{\frac{1}{650} \times 10^{-3} m} = \sin\theta_2 \cong 0.822$

$\Rightarrow 55.2° = \theta_2$

Thus, the angular separation between the second and first orders is 30.9°.

☐☐ **For the previous problem, how many complete orders are observed?**

The maximum angle θ in the expression $m\lambda = d\sin\theta$ is 90°. Thus,

$m(632 \times 10^{-9} m) = \left(\dfrac{1}{650} \times 10^{-3} m\right)(\sin 90°)$

$\Rightarrow$ m = 2.4, so that there are 2 maxima on either side of the central bright fringe.

☐☐ **Yellow light of wavelength 590 nm falls on a diffraction grating and the third order maximum occurs at 28°. Find (a) the separation between the slits of the grating, and (b) the number of lines per centimeter for this diffraction grating.**

(a) Here, $m\lambda = d\sin\theta$
    $(3)(590 \times 10^{-9} m) = d\sin 28°$, so that $d = 3.77 \times 10^{-6} m = 3.77 \times 10^{-4}$ cm.

(b) If 1 line = 3.77×10^{-4} cm, then 1 cm = 2650 lines or the grating has 2650 lines/cm.

☐☐ **Two stars, equally distant, are 5.7×10^{17} m from earth. How close could these stars be and be resolved by a telescope that has an objective of 1.02 m in light of wavelength of 632 nm? (Ignore atmospheric turbulence.)**

Here, $\dfrac{1.22\lambda}{a} = \dfrac{s}{r} = \dfrac{1.22(632 \times 10^{-9} m)}{1.02 m} = \dfrac{s}{5.7 \times 10^{17} m}$

$s = 4.31 \times 10^{11}$ m

The stars could be no closer than 4.31×10^{11} m and be resolved with this telescope.

☐☐ **Monochromatic light shines through a single slit of width 6×10^{-4} m. On a screen that is 3.9 m away, the distance between the central bright fringe and the first dark fringe is 3.4 mm. Find the wavelength of the light.**

Here, for dark bands, $m\lambda = a\sin\theta = \dfrac{ay}{L}$.

Thus $(1)(\lambda) = \dfrac{(6 \times 10^{-4} m)(3.4 \times 10^{-3} m)}{3.9 m}$

$\lambda = 5.23 \times 10^{-7} m$

The wavelength of the light is 523 nm.

☐☐ The distance between the lines of a diffraction grating is 2.3 x 10⁻⁶ m. How wide is the first order spectrum on a screen that is 3.5 m away if visible light is incident upon the grating?

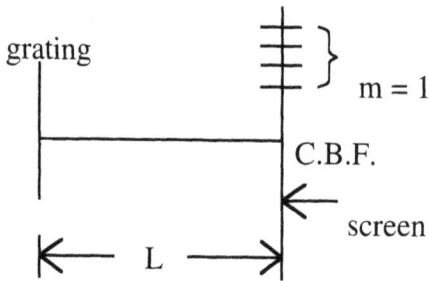

Visible light is light with wavelenghts from 400 - 700 nm. Thus for first order, the diffraction grating equation $m\lambda = d\sin\theta = \frac{yd}{L}$ or $y = \frac{m\lambda L}{d}$.

For the 400 nm light, $y = \frac{(1)(400 \times 10^{-9} m)(3.5 m)}{2.3 \times 10^{-6} m}$

$y_{400nm} = 0.609$ m.

For the 700 nm light, $y = \frac{(1)(700 \times 10^{-9} m)(3.5 m)}{2.3 \times 10^{-6} m}$

$y_{700nm} = 1.07$ m.

Thus, on the screen the width of the first order spectrum is $y_{700nm} - y_{400nm} = 0.461$ m.

☐☐ How far away could the taillights of a car be and still be seen by the eye (diameter of pupils is 7 mm, and wavelength of the light is 660 nm) as two distinct objects if (a) no attempt is made to compensate for the change of wavelength in the fluid of the eye, and (b) if the fluid in the eye has an index of refraction of 1.36? Assume the taillights are separated by 1.3 m.

(a) From Rayleigh's criterion,

$\frac{1.22\lambda}{a} = \frac{s}{r} \Rightarrow \frac{(1.22)(660 \times 10^{-9} m)}{7 \times 10^{-3} m} = \frac{1.3 m}{r} \Rightarrow r = 11,301 m$.

(b) Since $\lambda_{mat} = \frac{\lambda_{air}}{n} \Rightarrow \lambda_{mat} = \frac{660 nm}{1.36} = 485 nm$

Thus, $\frac{1.22(485 \times 10^{-9} m)}{7 \times 10^{-3} m} = \frac{1.3}{r} \Rightarrow r \cong 15,400 m$.

Hence, by compensating for the fact that the wavelength of the light within the eye is different than the wavelength in air, the eye would be able to resolve the taillights at 15.4 km.

❏❏ What would the diameter of the camera lens have to be to resolve the edges of a pack of cigarettes (assume the pack is 8 cm square) from a height of 100,000 ft (3.05×10^4 m)? (Assume that air turbulence can be ignored, and that the wavelength of the light is 550 nm.)

Here, from Rayleigh's criterion, $\dfrac{1.22\lambda}{a} = \dfrac{s}{r} \Rightarrow \dfrac{(1.22)(550 \times 10^{-9}\,m)}{a} = \dfrac{8 \times 10^{-2}\,m}{3.05 \times 10^4\,m}$

a = 0.256 m = 25.6 cm.

❏❏ A single slit has a width of 5×10^{-6} m. How many dark fringes would be formed on a screen 3.5 m away if the light from a He-Ne laser ($\lambda = 632$ nm) is incident upon the slit?)

For the dark bands of a single slit, $m\lambda = a\sin\theta$. The maximum angle is 90°.

Thus, $m = \dfrac{a\sin\theta}{\lambda} = \dfrac{(5 \times 10^{-6}\,m)(\sin 90°)}{550 \times 10^{-9}\,m} = 9$. Thus, there would be 9 dark bands formed on either side of the central bright fringe.

❏❏ Light that has wavelength of 650 nm is incident upon a single slit that has a width of .2 mm. Find the angular deflection for (a) the first minimum, and (b) the second minimum.

Since for a single slit the dark bands are given by $m\lambda = a\sin\theta = \dfrac{ay}{L}$,

(a) the first minimum angular deflection is given by

$\sin\theta = \dfrac{m\lambda}{a} = \dfrac{(1)(650 \times 10^{-9}\,m)}{2 \times 10^{-4}\,m}$

Thus, θ = 0.186° for the angular deviation of the first minimum.

(b) for the second minimum, $\sin\theta = \dfrac{(2)(650 \times 10^{-9}\,m)}{2 \times 10^{-4}\,m} = 6.5 \times 10^{-3}$

sin θ = 6.5 × 10⁻³

θ = 0.372° for the angular deviation of the second minimum.

❏❏ For the previous problem, what is the separation distance between the central bright fringe and the first maximum if the pattern is displayed on a screen 2 m from the single slit?

The bright fringes are given approximately by $\left(m + \dfrac{1}{2}\right)\lambda = a\sin\theta = \dfrac{ay}{L}$.

Thus, for the first bright fringe after the central bright fringe,

$y = \dfrac{L\left(m + \dfrac{1}{2}\right)\lambda}{a} = \dfrac{(2m)\left(\dfrac{1}{2}\right)(650 \times 10^{-9}\,m)}{2 \times 10^{-4}\,m}$

y = 3.25 × 10⁻³ m.

☐☐ **Microwaves are incident upon a single slit 11 cm wide, and the angular deflection to the second minimum is 36°. Find the wavelength of the microwaves.**

For a single slit, the minima are given by $m\lambda = a\sin\theta = \dfrac{ay}{L}$.

Thus, $(2)\lambda = (0.11\text{m})\sin 36°$

$\Rightarrow \lambda = 3.23 \times 10^{-2}$ m = 3.23 cm.

www.ingramcontent.com/pod-product-compliance
Ingram Content Group UK Ltd.
Pitfield, Milton Keynes, MK11 3LW, UK
UKHW051301180426
11947UKWH00020B/1833